Springer Theses

Recognizing Outstanding Ph.D. Research

Aims and Scope

The series "Springer Theses" brings together a selection of the very best Ph.D. theses from around the world and across the physical sciences. Nominated and endorsed by two recognized specialists, each published volume has been selected for its scientific excellence and the high impact of its contents for the pertinent field of research. For greater accessibility to non-specialists, the published versions include an extended introduction, as well as a foreword by the student's supervisor explaining the special relevance of the work for the field. As a whole, the series will provide a valuable resource both for newcomers to the research fields described, and for other scientists seeking detailed background information on special questions. Finally, it provides an accredited documentation of the valuable contributions made by today's younger generation of scientists.

Theses are accepted into the series by invited nomination only and must fulfill all of the following criteria

- They must be written in good English.
- The topic should fall within the confines of Chemistry, Physics, Earth Sciences, Engineering and related interdisciplinary fields such as Materials, Nanoscience, Chemical Engineering, Complex Systems and Biophysics.
- The work reported in the thesis must represent a significant scientific advance.
- If the thesis includes previously published material, permission to reproduce this must be gained from the respective copyright holder.
- They must have been examined and passed during the 12 months prior to nomination.
- Each thesis should include a foreword by the supervisor outlining the significance of its content.
- The theses should have a clearly defined structure including an introduction accessible to scientists not expert in that particular field.

More information about this series at http://www.springer.com/series/8790

Jungmin Seo

Development of Implantable Electronics as Novel Approaches to Obstructive Sleep Apnea

Doctoral thesis accepted by
Seoul National University, Korea

Springer

Author
Dr. Jungmin Seo
Seoul National University
Seoul, Korea (Republic of)

Supervisor
Prof. Sung June Kim
Seoul National University
Seoul, Korea (Republic of)

ISSN 2190-5053 ISSN 2190-5061 (electronic)
Springer Theses
ISBN 978-981-15-8326-1 ISBN 978-981-15-8327-8 (eBook)
https://doi.org/10.1007/978-981-15-8327-8

This Springer imprint is published by the registered company Springer Nature Singapore Pte Ltd.
The registered company address is: 152 Beach Road, #21-01/04 Gateway East, Singapore 189721, Singapore

To my mother and father, for their devotion and endless supports

And to my wife Minyoung, for her patience and sincere love

Supervisor's Foreword

Sleep is a precious gift offered each night to restore and rebalance exhausted mind and body. We need it to achieve a successful daily life. Unfortunately, however, modern-day humans suffer from various sleep disorders. Obstructive sleep apnea (OSA) is one of the sleep disorders, but it is often considered as mere discomfort and overlooked as snoring. Untreated, it may not only disrupt the daily life causing excessive daytime sleepiness, but also inflict serious complications such as insulin resistance, vascular diseases, stroke, and even death. Many therapies have been developed for treating OSA, in the past, including the Continuous Positive Airway Pressure (CPAP), the Maxilla-Mandibular Advancement (MMA), and the Uvulo Palato Pharyngo Plasty (UPPP). Many patients find these therapies helpful, but there still seems to be a need for improvement.

The work presented in this thesis is associated with a method based on neuromuscular electrical stimulation. There are two approaches suggested by the author, each of which targets major obstruction sites in the upper airway: the retrolingual and the retropalatal region, respectively. In the first, a cuff electrode system is designed and demonstrated for stimulating the hypoglossal nerve to achieve the upper airway patency on the retrolingual region. Two rare earth magnets are embedded in a substrate, allowing the cuff electrode to be repeatedly installed.

In the second, a palatal implant system is devised, to target the retropalatal opening by applying stimuli directly to the soft palate. Comprised of a palatal implant, an intraoral device, and a remote controller, the palatal implant system is featured with wireless control of the palatal implant via Zigbee communication and wireless powering of the implant via inductive link. The implant is carefully designed to be easily inserted into the soft palate and to provide bipolar current stimuli to the target area. The intraoral device, shaped as a mouthpiece, provides power to the implant, and plays a role in relaying the implant and the remote controller. Features of the systems are intensively verified and demonstrated in a bench-top, in vitro, and in vivo tests step by step.

It is my hope that the contents of this thesis, featuring many novel ideas, can be useful to further develop methods, eventually a product, to help millions who suffer from this sleep breathing issue.

Seoul, Korea (Republic of)
August 2020

Prof. Sung June Kim

Preface

Obstructive sleep apnea (OSA) is a disorder with high prevalence among the population, affecting 13% men and 6% women in North America. OSA is often characterized by mere nighttime discomfort, but its chronic state may involve sequelae such as insulin resistance, vascular diseases, and impaired quality of life due to excessive sleepiness. Although various therapies are available for OSA, such as applying continuous positive airway pressure, using oral appliances, and undergoing surgical operations, they are not suitable for some patients owing to low effectiveness, low adherence, or high invasiveness. Therefore, it is necessary to develop breakthrough therapies, which can overcome the limitations of conventional therapies.

Electrical engineering is a synergetic area, which can converge with biology because nerves and muscles can be easily controlled using electricity. Neuroprosthetics is a research field that leverages this synergy. With common devices such as pacemakers, cochlear implants, deep brain stimulators, and artificial retinae, neuroprosthetics allows treatment of incurable diseases or restores senses. As many patients suffering from OSA cannot find an absolute cure, this thesis aims to devise OSA treatments based on neuroprosthetics.

Two main ideas are proposed in this thesis as potential OSA treatments. First, a magnetic cuff system comprising nerve cuff electrodes embedded with magnets and a current pulse generator. The cuff electrode wraps the hypoglossal nerve using the magnetic force and stimulates the nerve with electrical pulses to achieve retrolingual opening. Second, a palatal implant system comprising an intraoral device, a palatal implant, and a remote controller. This system can be used to control the retropalatal region.

By following the course of design, device fabrication, and experimental assessment, readers can understand the design of implantable devices, their fabrication, and their evaluation. For instance, the selection of biocompatible materials, microfabrication, packaging, circuit design, coil antenna design, and application-specific integrated circuit design are explained in this thesis. Given its multidisciplinary nature, this thesis avoids profound and specialized theories and instead, provides wide knowledge about topics pertaining to electricity, chemistry, biology, and medicine. Therefore, not only readers from these fields but also any reader who is interested in implantable electronics is able to follow and understand the content in this thesis.

Chapter 1 provides a background on OSA and neuroprosthetics.

Chapter 2 deals with circuit design and fabrication of electrodes. A reader can learn important concepts for designing circuits and electrodes and follow descriptions of the fabrication process.

Chapter 3 presents experimental results of this thesis. Brief discussions and numerical values corresponding to the results are provided.

Chapter 4 provides discussions on selected topics. Future directions of work for further improving the developed devices are included.

Chapter 5 summarizes this thesis and provides conclusions.

In summary, the focus of this thesis is the design, fabrication, and evaluation of neuroprosthetic implantable devices and their application to OSA treatment. Electrical engineers, bioelectronic researchers, and readers from related fields can find useful content in this thesis.

Seoul, Korea (Republic of) Jungmin Seo

Notes

Some parts of this dissertation are extracted and adapted from the following publications which were published during the course of this study:

- Seo, J., Wee, J. H., Park, J. H., Park, P., Kim, J. W., & Kim, S. J. (2016). Nerve cuff electrode using embedded magnets and its application to hypoglossal nerve stimulation. Journal of neural engineering, 13(6), 066014.
- Seo, J., Kim, J. W., Cho, S. W., Shim, S., Choi, J. W., & Kim, S. J. (2018, July). Preliminary Study of Palatal Implant for Sleep Apnea Control. In 2018 40th Annual International Conference of the IEEE Engineering in Medicine and Biology Society (EMBC) (pp. 1498–1501). IEEE.
- Seo, J., Yun, S., Shim, S., Cho, S. W., Choi, J. W., Kim, J. W., & Kim, S. J. (2020). Palatal implant system can provide effective treatment for obstructive sleep apnea by recovering retropalatal patency. Journal of Neural Engineering, 17(2), 026017.

Acknowledgements

It may be only appropriate to express my gratitude herein to Professor Sung June Kim (Seoul National University). He gave me sincere guidance and teachings throughout my doctoral course. Undoubtedly, all my achievements and this thesis could not have been done without him.

I want to express my heartfelt thanks to Professor Jeong-Whun Kim (Seoul National University Bundang Hospital), who unsparingly taught me medical knowledge and gave me advice on papers. It was such a pleasure to experience engineering–medical convergence, which was invaluable and cannot be easily done.

I am also thankful to my colleagues, seniors, and juniors in the NanoBio-electronics & Systems Laboratory. Without them, the difficulties or limitations encountered during the research could not have been overcome.

I want to express my additional thanks to the staff in the EECS (Electrical Engineering and Computer Science), Seoul National University, and in Springer, particularly, Min A Kim, Kwang Suk Oh, Smith Chae, and Divya Meiyazhagan. They helped me a lot with consistent kindness and professional support to write the thesis.

Finally, I would like to thank my wife, Minyoung, for raising numerous questions to improve the thesis. Her different perspectives made me re-examine my understanding and strengthen my knowledge.

Contents

1 Introduction . . . 1
1.1 Obstructive Sleep Apnea (OSA) . . . 1
1.1.1 Pathogenesis of OSA . . . 2
1.1.2 Conventional Therapies . . . 2
1.2 Electrical Stimulation for OSA Treatment . . . 4
1.2.1 Neural and Neuromuscular Stimulation . . . 4
1.2.2 Hypoglossal Nerve Stimulation (HNS) . . . 5
1.2.3 Soft Palate Stimulation (SPS) . . . 6
1.3 Suggested Approaches . . . 6
1.3.1 Liquid Crystal Polymer (LCP) . . . 6
1.3.2 Magnetic Cuff System for HNS . . . 7
1.3.3 Palatal Implant System for SPS . . . 9
1.4 Objectives of This Dissertation . . . 10
References . . . 10

2 Methods . . . 15
2.1 Magnetic Cuff System for HNS . . . 15
2.1.1 Overview . . . 15
2.1.2 Magnet-Embedded Nerve Cuff Electrode . . . 15
2.1.3 External Pulse Generator . . . 18
2.1.4 Evaluations . . . 19
2.2 Palatal Implant System for SPS . . . 21
2.2.1 Pilot Study in Vivo . . . 21
2.2.2 Overview . . . 23
2.2.3 Palatal Implant . . . 24
2.2.4 Intra-Oral Device . . . 30
2.2.5 Evaluations . . . 38
References . . . 41

3 Results 43
3.1 Magnetic Cuff System for HNS 43
3.1.1 Fabricated System 43
3.1.2 Electrochemical Measurements in Vitro 43
3.1.3 Animal Testing in Vivo 45
3.2 Palatal Implant System for SPS 47
3.2.1 Pilot Study in Vivo 47
3.2.2 Fabricated System 48
3.2.3 Wireless Power Transmission 51
3.2.4 Wireless Data Transmission 52
3.2.5 Electrochemical Measurements in Vitro 53
3.2.6 Animal Testing in Vivo 58
References 60

4 Discussions 61
4.1 Magnetic Cuff System for HNS 61
4.1.1 Cuff-to-Nerve Diameter Ratio 61
4.1.2 Simulation Parameters 62
4.1.3 Extension to Multi-channels 62
4.2 Palatal Implant System for SPS 63
4.2.1 Reliability of Intra-Oral Device 63
4.2.2 Palatoglossus Coupling 64
4.2.3 Potential Developments 65
References 70

5 Conclusion 73

Author Biography 75

Abbreviations

ASIC	Application-Specific Integrated Circuit
CPAP	Continuous Positive Airway Pressure
EMG	Electromyogram
FEM	Finite Element Method
FFC	Flexible Flat Cable
HNS	Hypoglossal Nerve Stimulation
IC	Integrated Circuit
IOD	Intra Oral Device
LAUP	Laser-Assisted Uvuloplatoplasty (LAUP)
LCP	Liquid Crystal Polymer
MAD	Mandibular Advancement Device
MENCE	Magnet-Embedded Nerve Cuff Electrode
MMA	Maxilo-Mandibular Advancement
OSA	Obstructive Sleep Apnea
PCB	Printed Circuit Board
PWM	Pulse-Width Modulation
SPICE	Simulation Program with Integrated Circuit Emphasis
SPS	Soft Palate Stimulation
UPPP	Uvulopalatopharyngoplasty

Chapter 1
Introduction

Obstructive sleep apnea (OSA) is a high-prevalence disease that 13% of men and 6% of women suffer from in North America. OSA can be regarded as a mere discomfort; however, it may cause serious complications, such as insulin resistance, dyslipidemia, vascular disease, and even death. Therefore, it is very important to provide consistent and effective therapies to patients.

In this chapter, the pathogenesis and conventional therapies of OSA are introduced. Because OSA originates from two big muscles, the palate and the tongue, therapies for OSA are divided into methods that address tongue-related and palate-related OSA. Unconventional treatment methods, the neural stimulation and the neuromuscular stimulation, are introduced as the novel approaches to OSA; they feature the application of electrical stimuli to the target nerve or muscle.

1.1 Obstructive Sleep Apnea (OSA)

Obstructive sleep apnea (OSA) is becoming a common disease in the modern society; it has been reported that 13% of men and 6% of women suffer from OSA in North America [1]. Easily confused with snoring, OSA is often overlooked as a mere night-time discomfort. However, persistent OSA may lower the quality of life due to excessive sleepiness and chronic fatigue. Patients having an apnea-hypopnea index (AHI), the number of apnea or hypopnea occurred per night hour, of 15 or more may suffer critical sequelae, such as insulin resistance, dyslipidemia, vascular disease, and even death [2]. Therefore, it is important to provide effective therapies consistently to patients to treat OSA.

J. Seo, *Development of Implantable Electronics as Novel Approaches to Obstructive Sleep Apnea*, Springer Theses, https://doi.org/10.1007/978-981-15-8327-8_1

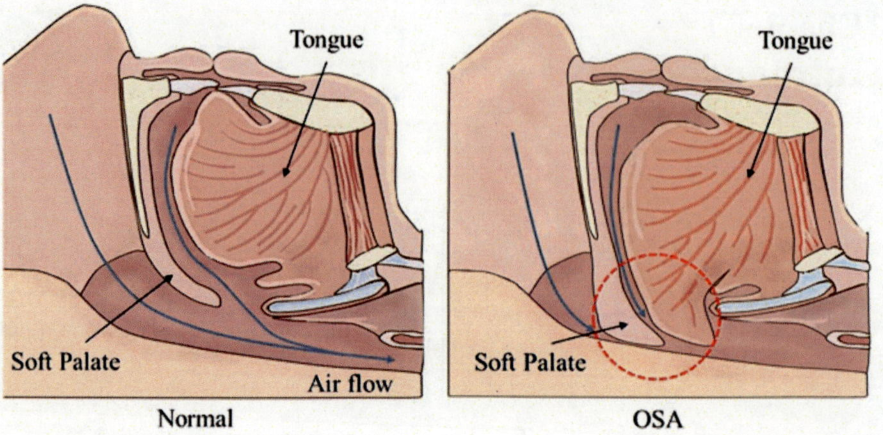

Fig. 1.1 Illustrations of the upper airway and its surrounding muscles in case of normal (left) and OSA (right)

1.1.1 Pathogenesis of OSA

Obstructive sleep apnea (OSA) is the most common type of sleep apnea. It is caused by partial or complete obstruction of the upper airway due to the surrounding muscles of the pharynx, regardless of respiratory effort of the lungs. There are forces that promote upper-airway collapse, such as intraluminal negative pressure, surface tension, and gravity, while tonic and phasic activity of muscles and wakefulness are factors that contribute to airway patency [3]. The forces applied to the surrounding muscles are balanced in normal individuals so that the upper airway does not collapse as shown in the left figure in Fig. 1.1. However, the upper airway collapse is mainly involved in the retrolingual or retropalatal region or both when the balance of the forces is broken, as shown in the right figure in Fig. 1.1. Therefore, for an effective treatment, it is necessary to first determine where the collapse has occurred and then to provide an appropriate therapy. Retrolingual opening can be accomplished by modifying the position of the tongue, and retropalatal opening can be accomplished by modifying the position or stiffness of the soft palate.

1.1.2 Conventional Therapies

As stated in 1.1.1, two major pharyngeal muscles, the tongue and the soft palate, are the main contributors to pharyngeal collapse. Continuous positive airway pressure (CPAP) helps maintaining patency at both the retrolingual and retropalatal regions by supplying pressurized air. Maxilomandibular advancement (MMA), mandibular advancement device (MAD), and oral appliances are therapies that open the retrolingual region by modifying the tongue position. Uvulopharyngopalatoplasty (UPPP), laser-assisted uvulopalatoplasty (LAUP), and pillar implants are therapies related to soft palate modification for retropalatal opening. Detailed explanations of the therapies are provided in 1.1.2.1 and 1.1.2.2.

1.1.2.1 Continuous Positive Airway Pressure

Continuous positive airway pressure (CPAP) is the most popular therapy owing to its non-invasiveness and high efficacy. CPAP delivers positively pressurized air into the upper airway via a nasal mask and a tube connected to a pump. Thus, the patency of the upper airway can be recovered by pushing out collapsed muscles from the upper airway [4]. CPAP has been reported to be very effective if correctly applied with an acceptable usage rate, over 4 h per night [5]. However, it also has been reported that under 50% of patients use CPAP over 4 h per night after 6 months, and the percentage of patients drops to 17% after 5 years, due to discomfort and inconvenience in using CPAP [6, 7]. This poor adherence lowers the effectiveness of CPAP therapy, considering that the effectiveness of the therapy can be expressed in the product of efficacy and adherence of patients to the therapy.

1.1.2.2 Therapies for Tongue-Related OSA

Therapies of the tongue-related OSA focus on recovering retrolingual patency by modifying position of the tongue. Maxilomandibular advancement (MMA), an orthognathic surgery that includes advancement of the mandible, is a surgical operation to the tongue-related OSA patients. Often performed simultaneously with Genioglossus advancement (GA), MMA has been demonstrated to be one of the most effective therapy for treating sleep apnea having AHI reduction of 87% [8]. However, MMA has been reported to accompany negative functional changes as well as cosmetic changes after the operation. Significant percentages (20–50%) of patients presented disorders related to the masticatory apparatus, gum pain, dental wear, and long-lasting postoperative sensory disorders. Furthermore, 89.4% of patients reported changes in their facial appearances and 18.3% were unsatisfied with their changed faces [9].

An oral appliance is a mouth-equipping device during sleep to advance position of the mandible and thereby, advances the tongue as well. As the appliance is non-invasive and easily applicable, it is recommended to patients who failed to adhere to the CPAP treatment. The average efficacy of the oral appliances is reported to be 45%, which is far lower than that of CPAP, 87%. However, the average adherence is 95%, while the adherence remains under 50% for CPAP treatment [10, 11]. The excellent adherence of the oral appliance compensates the poor efficacy of the oral appliance treatment, making the therapy comparably effective to the CPAP.

1.1.2.3 Therapies for Palate-Related OSA

Uvulopharyngopalatoplasty (UPPP) is a commonly performed surgical procedure for retropalatal opening. UPPP involves removal and rearrangement of the tonsils, adenoids, uvula, soft palate, and pharynx to increase the upper airway size and thus prevent tissue collapse [12]. Laser-assisted uvulopalatoplasty (LAUP), which is similar to UPPP regarding the removal of upper airway tissues, utilizes a laser to vaporize the uvula or a specified portion of tissues under local anesthesia. Both UPPP and LAUP are direct methods to remove excessive tissues, but there are a few side-effects reported, namely, respiratory compromise, bleeding, and death in 0 to 16% of patients [13–16]. Persistent side-effects were also reported in a mean of 59%, including difficulty in swallowing, globus sensation, voice changes, smell disturbances, and velopharyngeal stenosis after LAUP [17–22]. Regarding efficacy, a 20% decrease in the mean AHI has been reported for LAUP, and 24 to 33% for UPPP, which are not satisfactory results considering the possible side-effects [23, 24].

The pillar implant system (Medtronic, USA), introduced as a simple treatment for snoring in 2003, enhances the stiffness of the soft palate through a mechanical support [25, 26]. Three pillars are implanted into the soft palate to support it; submucosal thickening is achieved through the creation of fibrotic capsules around the implants. The easy implantation procedure under local anesthesia in an office environment is a significant advantage of the pillar implant system. However, its efficacy has been reported to be limited [27, 28]. According to Server et al., no statistically significant differences on the Epwort sleepiness scale (ESS) and AHI were detected, while the average of the visual analog scale (VAS) was decreased by only 18.1% [29]. The low efficacy seems to be due to the pillars' passive support rather than active contraction of the muscle.

1.2 Electrical Stimulation for OSA Treatment

As mentioned in 1.1.2, the conventional therapies all have their own shortcomings. Therefore, there is a necessity to develop a new treatment method that overcomes such drawbacks. Here, we suggest electrical stimulation as an alternative treatment method for OSA.

1.2.1 Neural and Neuromuscular Stimulation

Nerves represent their firing data in polarized voltages, called action potentials, across their membranes. When the firing data are delivered to the innervating sensory or motor cells, their corresponding functions appear. Therefore, the functions can be modulated by forcing the polarization of the membrane potential using electric pulses.

Many neuro-prosthetic devices have been developed based on the principle of neuromodulation. Cochlear implants are appraised to be one of the most successful neural prostheses developed so far [30–33]. The implant substitutes impaired hair cells, which play role in transducing sounds to electric pulses, by applying electric stimulation to the cochlea's hearing nerve. A retinal implant is based on a similar strategy but to stimulate the optic nerve [34–36]. There are further neural prostheses based on spinal cord stimulation, deep brain stimulation, and hypoglossal nerve stimulation to modulate certain neural functions [37–41].

Neuromuscular stimulation induces muscle contraction using electric pulses similar to that used in neural prostheses. The pulses mimic action potentials from the central nerve system, causing the muscles to contract by stimulating either innervating nerves or neuromuscular junctions. Neuromuscular stimulation has been used for clinical applications to provide either a functional or therapeutic effect. Functional neuromuscular stimulation includes the activation of paralyzed muscles, support of standing or ambulatory activities, or control of respiration and bladder function [42–46]. The therapeutic effects of neuromuscular stimulation include motor relearning, which refers to the recovery of motor skills that have been lost due to damage to the central nerve system [47, 48].

1.2.2 Hypoglossal Nerve Stimulation (HNS)

With the advances of neural prostheses, a hypoglossal nerve stimulation technique has been developed to treat OSA by stimulating the innervating nerve of the genioglossus, the extrinsic muscle of the tongue, to accomplish retrolingual opening. In 1980, Brouillette et al. first found that control of the genioglossus, the extrinsic muscle of the tongue, may contribute to inspiratory activity [49]. Inspired by the possibility, hypoglossal nerve stimulation has been investigated since the late 90 s, considering the fact that the hypoglossal nerve innervates the genioglossus [50–53]. Recently, a clinical study using an implantable hypoglossal nerve stimulator (Inspire Medical Systems, USA) was reported in the *New England Journal of Medicine*. The results showed that the median AHI score at 12 months decreased by 68%, from 29.3 events per hour to 9.0 events per hour with the overall rate of serious adverse events less than 2%, through a 12-month follow-up on patients using the stimulator [38]. Furthermore, an interesting phenomenon has been found through other clinical studies that hypoglossal nerve stimulation also induces retropalatal opening although the stimulation originally targets the retrolingual opening. This phenomenon, called palatoglossus coupling, is due to the palatoglossus muscle that connects the anterior pillar from the uvula and the sides of the tongue. The elicited coupling was reported to exist in nearly 80% of patients implying the possibility of multi-region opening from single stimulation on that population [54].

1.2.3 Soft Palate Stimulation (SPS)

The palatal muscles play an important role in upper-airway patency considering that pharyngeal collapse often occurs in the velopharyngeal region [55]. The palatal muscles consist of five major muscles parts: the muscles controlling the stiffness (*tensor veli palatini*) and position (*levator veli palatini*) of the palate, tongue (palatoglossus), pharynx (palatopharyngeus), and the shape of the uvula (musculus uvulae).

Despite the importance of the palatal muscles in upper-airway obstruction, few trials have been conducted to control those muscles because they are less accessible than the genioglossus. In 1996, Schwartz conducted a study to determine whether electrical stimulation of the soft palate has an effect on ceasing either snoring or OSA [56]. Seven men between the ages of 35 and 49 were selected from a pool of patients who were diagnosed with OSA with snoring. The oral appliances used in the study had two wire electrodes, and voltage pulses were applied to the soft palate via the electrodes. The result showed that 25% of patients experienced cessation of snoring and varied effects on OSA with voltage amplitudes in the range of 2.5–9.0 V, without arousal. A more sophisticated experiment was conducted by selective stimulation of the *tensor veli palatini* muscle while the upper airway patency was monitored represented in critical pressure. There was a significant reduction in the critical pressure, indicating the therapeutic effect of palatal stimulation for OSA treatment [57].

1.3 Suggested Approaches

Here we suggest two novel approaches for OSA treatments as alternatives to the conventional therapies. In this thesis, two implantable electronics are suggested; one is for retrolingual opening, and the other is for retropalatal opening. First, an introduction is provided about the biocompatible polymer used in the devices, and then the devices are briefly described.

1.3.1 Liquid Crystal Polymer (LCP)

The suggested devices are implantable electronics that should remain in the body for a long time. Therefore, it is important to use a biocompatible and hermetic material that allows the device to remain safe and stable in the body. Metals are widely used as packaging materials for implantable devices. Among them, titanium has been a very popular material because it provides long-term biocompatibility and is resistant to corrosion by the body fluid when alloyed with other elements. However, there are a few problems of using metal packages. First, metal packages cannot be fabricated as light and small as other packages, specifically polymers. The bulky and heavy package are usually placed in the chest where there is sufficient space. Therefore, long connection lines are required to the stimulation electrodes, which are far from the package. This involves an invasive operation to place the connection lines under the skin. Furthermore, because a metal package cannot be monolithically integrated with electrodes, there are hermeticity issues on the feed-throughs. The hetero-junction of the lead-lines and the package increases the possibility of device failure at the feed-throughs.

Polymers are another material options for packaging implantable devices. Polyimide, parylene-C, silicone elastomer, and liquid crystal polymer (LCP) are such polymers, all of which are frequently utilized as packaging materials. Among them, LCP has extremely low water absorption under 0.04%, which 100 times lower than that of polyimide at maximum [58]. Due to its high hermeticity in comparison to other polymers, LCP is regarded as a suitable material for polymer packaging. Another advantage of using LCP is that the entire implantable device, including the electrodes, connection lines, feed-throughs, and pulse generators, can be fabricated and packaged monolithically. Previous works on LCP-based retinal implant devices, cochlear implants, and depth probes have been reported along with their reliability demonstrations and long-term evaluations [59–61]. In addition, LCP is compatible with various conventional manufacturing processes, from UV laser machining to the microfabrication processes of photolithography, metal deposition, and wet/dry etching in a wafer unit. The glass-transition temperature and melting temperature are relatively low, in the range of 150–335 °C; therefore, a multi-layered device can be easily fabricated through a lamination procedure. Also, Young's modulus of LCP complies with those of other flexible polymers, such as polyimide or parylene-C, and the rigidity of the device can be easily controlled by varying its thickness [62, 63]. Finally, LCP has good high-frequency properties, a relative permittivity of 2.9, and a tangent loss of 0.0022 at 25 GHz [64]. Table 1.1 presents a comparison of LCP and other polymers.

1.3.2 Magnetic Cuff System for HNS

The hypoglossal nerve is the twelfth cranial nerve, which innervates the genioglossus, an extrinsic muscle of the tongue. In humans and mammals, the terminating branches of the hypoglossal nerve are connected to both the retractors and protrusors of the tongue [65]. The hyoglossus muscle, the retractor of the tongue, comprises between two and five branches, which are laterally divided from the hypoglossal nerve; therefore, it is difficult to distinguish them from each other [66]. Furthermore, even a single excited branch of the hyoglossus muscle can induce retraction of the tongue [54]. Accordingly, correct placement of the electrode is crucial for successful therapy [67].

There have been numerous trials to achieve effective interfacing to nerves. Among them, a cuff electrode, consisting of metal sites on a flexible insulating substrate, was developed and proved to be advantageous over interfacing to nerves. A cuff electrode

Table 1.1 Comparison table of biocompatible polymers on physical properties

	LCP (Vecstar)	Polyiinide (PI2525)	PDMS
Melting temp. (°C)	280–335	>400	200–250
Tensile strength (MPa)	270–500	128	2.24
Young's modulus (GPa)	2–10	1.8–15	0.1–0.87
Water absorption (%)	<0.04	0.24–4	0.1–1.3
Dielectric constant (@lMHz)	2.9	3.3	2.6

can conformably adhere to the cylindrical nerve owing to its flexibility. This provides features that are not found in other types of electrodes, including a low threshold level, a reduced probability of lead failure, and the capability of use regardless of the surrounding muscle length [68].

Most of the developed cuff electrodes are split-cylinder type, and they require a closing method at the time of installation. Suturing, the most popular closing method in conventional medical surgery, has been popular for cuff electrodes as well [69, 70]. A surgeon ties and stiches a seam of the cuff with a suture after wrapping the nerve. It is a simple and powerful closing method, but it is labor-intensive in a situation where repetitive installation is required. Also, if tightened excessively, the nerve can be damaged and impaired due to decreased blood flow.

A closing method using adhesives has been also developed to reduce labor during installation. There have been trials to use a small bead of an adhesive using silicone or any biocompatible wax at the seam of the cuff. A shape memory alloy was adopted for another closing method. Three sets of an armature consisting of a nickel-titanium wire split ring were positioned on the silicone rubber cuff.

A few ideas to make a self-spiral cuff have been reported [68, 71–73]. The self-spiraling cuff electrodes do not require closing methods because they curl themselves in a natural state. Several groups have tried to utilize a shape memory alloy for the cuff. The temperature-dependent characteristic of the alloy allows the cuff to be opened at a low temperature and closed at body temperature [74, 75]. Other trials for the self-spiral cuff include a method to bond two sheets of silicone rubber together having different natural-resting lengths, and to deform a cuff into a curled shape at 350 °C under a nitrogen atmosphere [68, 70].

So far, many methods to close the cuff have been suggested, and they all have their own contribution to closing. However, none of them are appropriate for hypoglossal nerve stimulation, where several attempts are required for correct installation. Near the hypoglossal nerve, there are confusable other nerves including key retractors, such as the styloglossus and the hyoglossus, which must not be activated to achieve upper-airway opening [67]. Therefore, the development of a cuff electrode with a repeatable and ensured closing feature is necessary for the application.

Here, we propose a magnet-embedded neural cuff electrode (MENCE), which enables repeatable closure using magnetic force. Two rare-earth magnets are embedded at each side of the cuff and fully encapsulated by LCP. MENCE with two stimulation sites was produced by a microfabrication process that included photolithography, E-gun evaporation, and wet/dry etching. An external pulse generator was designed and fabricated using an ASIC to comprise a stimulation system. MENCE was evaluated in vivo as well as in vitro, to demonstrate its feasibility. In vitro evaluations of the electrochemical impedance spectrum (EIS) and cyclic voltammetry (CV) were conducted under condition of PBS solution. After the in vitro evaluations, MENCE was installed on the hypoglossal nerve of a rabbit and connected to a pulse generator to deliver current stimuli to the electrode. During the stimulation, tongue movement was visually observed, and the upper airway was recorded by C-arm fluoroscopy.

1.3.3 Palatal Implant System for SPS

The effect of soft palate stimulation was partly demonstrated on ceasing snoring in previous studies [56, 57]. To clearly demonstrate the effect of the electrical stimulation on the soft palate, a pilot study was conducted by applying stimuli to the soft palate of a rabbit with an observation of the upper airway through C-arm fluoroscopy.

Based on the result of the pilot study, a palatal implant system was developed. The system consists of two main parts, an implant that electrically stimulates the soft palate, and an intra-oral device (IOD) that transmits power and data to the implant. There is an auxiliary device called a wireless controller, which provides a user interface to send stimulation parameter settings to the IOD via ZigBee. The implant is fabricated by laminating multiple LCP layers, comprising coil, electrode, circuit, and adhesion layers. The IOD includes a rechargeable battery, ZigBee receiver, amplifier, and transmitting coil antenna, all of which are integrated and packaged by a silicone elastomer. The inductive link was evaluated in terms of power transmission and data reliability on telemetry distances ranging from 2 to 8 mm. Further assessments were conducted in vitro to evaluate the feasibility of the electrode. Electrochemical impedance spectroscopy (EIS), cyclic voltammetry (CV), and voltage transient measurements were performed by a three-electrode cell in a PBS solution. Then the system was applied to a rabbit, and the response of the soft palate was observed by C-ram fluoroscopy. Finally, reliability, genioglossus coupling, and possible developments are discussed.

Table 1.2 Objectives and brief descriptions of this dissertation

Goal	Design	Fabrication	Evaluation
Development of magnetic cuff system–for **retrolingual** opening	– For hypoglossal nerve stimulation – Repeatedly installable cuff electrode using magnetic force	– Photolithography, E- gun evaporation, diy/wet etching – UV Laser machining – Thermal lamination	– Electro-impedance spectroscopy – Cyclic Voltammetry – In Vivo Stimulation
Development of palatal implant system for **retropalatal** opening	– For soft palate stimulation – Palatal implant with intra-oral device	– Fabrications above – Component mounting – PCB design and fabrication – Encapsulation using silicone elastomer	– Telemetry Reliability – Electro-impedance spectroscopy – Cyclic Voltammetry – Voltage Transient – In Vivo Stimulation

1.4 Objectives of This Dissertation

The major objective of this study was to develop neural prostheses for OSA treatment. Because OSA involves either retrolingual or retropalatal collapse, two implantable electronics were devised and developed to address each of these problems (Table 1.2).

First, a magnetic cuff system was developed for hypoglossal nerve stimulation. Hypoglossal nerve stimulation is an effective technique for retrolingual opening, but it requires iterative trials to find the correct location for installation. The developed cuff electrode can play a role in that it allows easy repetitive installation and removal using magnetic forces.

For retropalatal opening, a palatal implant system is firstly proposed. Despite the development of hypoglossal nerve stimulation, some patients still suffer from retropalatal collapse. The problem can be solved by contracting the soft palate by the application of electric stimuli. The system can help palate-related patients who cannot benefit from the conventional therapies.

References

1. Peppard PE, Young T, Barnet JH, Palta M, Hagen EW, Hla KM (2013) Increased prevalence of sleep-disordered breathing in adults. Am J Epidemiol 177(9):1006–1014. https://doi.org/10.1093/aje/kws342
2. Lattimore JD, Celermajer DS, Wilcox I (2003) Obstructive sleep apnea and cardiovascular disease. J Am Coll Cardiol 41(9):1429–1437. https://doi.org/10.1016/s0735-1097(03)00184-0
3. Pengo MF, Steier J (2015) Emerging technology: electrical stimulation in obstructive sleep apnoea. J Thorac Dis 7(8):1286–1297. https://doi.org/10.3978/j.issn.2072-1439.2014.04.04
4. Sutherland K, Phillips C, Cistulli PJJoDSM (2015) Efficacy versus effectiveness in the treatment of obstructive sleep apnea: CPAP and oral appliances 2 (4):175–181

5. Kribbs NB, Pack AI, Kline LR, Smith PL, Schwartz AR, Schubert NM, Redline S, Henry JN, Getsy JE, Dinges DF (1993) Objective measurement of patterns of nasal CPAP use by patients with obstructive sleep apnea. J Am Rev Respir Dis 147(4):887–895
6. Bartlett D, Wong K, Richards D, Moy E, Espie CA, Cistulli PA, Grunstein R (2013) Increasing adherence to obstructive sleep apnea treatment with a group social cognitive therapy treatment intervention: a randomized trial. Sleep 36(11):1647–1654. https://doi.org/10.5665/sleep.3118
7. Weaver TE, Sawyer A (2009) Management of obstructive sleep apnea by continuous positive airway pressure. Oral Maxillofac Surg Clin North Am 21(4):403–412. https://doi.org/10.1016/j.coms.2009.08.001
8. Caples SM, Rowley JA, Prinsell JR, Pallanch JF, Elamin MB, Katz SG, Harwick JD (2010) Surgical modifications of the upper airway for obstructive sleep apnea in adults: a systematic review and meta-analysis. Sleep 33(10):1396–1407. https://doi.org/10.1093/sleep/33.10.1396
9. Blumen MB, Buchet I, Meulien P, Hausser Hauw C, Neveu H, Chabolle F (2009) Complications/adverse effects of maxillomandibular advancement for the treatment of OSA in regard to outcome. Otolaryngol Head Neck Surg 141(5):591–597. https://doi.org/10.1016/j.otohns.2009.08.025
10. Sutherland K, Phillips C, Cistulli P (2015) Efficacy versus effectiveness in the treatment of obstructive sleep apnea: CPAP and oral appliances. J Dent Sleep Med 2(4):175–181
11. Phillips CL, Grunstein RR, Darendeliler MA, Mihailidou AS, Srinivasan VK, Yee BJ, Marks GB, Cistulli PA (2013) Health outcomes of continuous positive airway pressure versus oral appliance treatment for obstructive sleep apnea: a randomized controlled trial. Am J Respir Crit Care Med 187(8):879–887. https://doi.org/10.1164/rccm.201212-2223OC
12. Labra A, Huerta-Delgado AD, Gutierrez-Sanchez C, Cordero-Chacon SA, Basurto-Madero P (2008) Uvulopalatopharyngoplasty and uvulopalatal flap for the treatment of snoring: technique to avoid complications. J Otolaryngol Head Neck Surg 37(2):256–259
13. Kezirian EJ, Weaver EM, Yueh B, Deyo RA, Khuri SF, Daley J, Henderson W (2004) Incidence of serious complications after uvulopalatopharyngoplasty. Laryngoscope 114(3):450–453. https://doi.org/10.1097/00005537-200403000-00012
14. Esclamado RM, Glenn MG, McCulloch TM, Cummings CW (1989) Perioperative complications and risk factors in the surgical treatment of obstructive sleep apnea syndrome. Laryngoscope 99(11):1125–1129. https://doi.org/10.1288/00005537-198911000-00004
15. Harmon JD, Morgan W, Chaudhary B (1989) Sleep apnea: morbidity and mortality of surgical treatment. South Med J 82(2):161–164
16. Riley RW, Powell NB, Guilleminault C, Pelayo R, Troell RJ, Li KK (1997) Obstructive sleep apnea surgery: risk management and complications. Otolaryngol Head Neck Surg 117(6):648–652. https://doi.org/10.1016/s0194-5998(97)70047-0
17. Ferguson KA, Heighway K, Ruby RR (2003) A randomized trial of laser-assisted uvulopalatoplasty in the treatment of mild obstructive sleep apnea. Am J Respir Crit Care Med 167(1):15–19. https://doi.org/10.1164/rccm.2108050
18. Larrosa F, Hernandez L, Morello A, Ballester E, Quinto L, Montserrat J (2004) Laser-assisted uvulopalatoplasty for snoring: does it meet the expectations? Eur Respir J 24(1):66–70
19. Walker RP, Gopalsami C (1996) Laser-assisted uvulopalatoplasty: postoperative complications. Laryngoscope 106(7):834–838. https://doi.org/10.1097/00005537-199607000-00009
20. Carenfelt C, Haraldsson PO (1993) Frequency of complications after uvulopalatopharyngoplasty. Lancet 341(8842):437. https://doi.org/10.1016/0140-6736(93)93030-5
21. Berger G, Stein G, Ophir D, Finkelstein Y (2003) Is there a better way to do laser-assisted uvulopalatoplasty? J Arch Otolaryngol Head Neck Surgery 129(4):447–453
22. Pavelec V, Polenik PJTL (2006) Use of Er, Cr: YSGG versus standard lasers in laser assisted uvulopalatoplasty for treatment of snoring 116 (8):1512–1516
23. Khan A, Ramar K, Maddirala S, Friedman O, Pallanch JF, Olson EJ Uvulopalatopharyngoplasty in the management of obstructive sleep apnea: the mayo clinic experience. In: Mayo Clinic Proceedings, 2009. vol 9. Elsevier, pp 795–800
24. Franklin KA, Anttila H, Axelsson S, Gislason T, Maasilta P, Myhre KI, Rehnqvist N (2009) Effects and side-effects of surgery for snoring and obstructive sleep apnea–a systematic review. Sleep 32(1):27–36

25. Ho WK, Wei WI, Chung KF (2004) Managing disturbing snoring with palatal implants: a pilot study. Arch Otolaryngol Head Neck Surg 130(6):753–758. https://doi.org/10.1001/archotol.130.6.753
26. Nordgård S, Wormdal K, Bugten V, Stene BK, Skjøstad KW (2004) Palatal implants: a new method for the treatment of snoring. J Acta oto-laryngologica 124(8):970–975
27. Gillespie MB, Smith JE, Clarke J, Nguyen SA (2009) Effectiveness of pillar palatal implants for snoring management. Otolaryngol Head Neck Surg 140(3):363–368. https://doi.org/10.1016/j.otohns.2008.12.027
28. Friedman M, Schalch P, Joseph NJ (2006) Palatal stiffening after failed uvulopalatopharyngoplasty with the Pillar Implant System. Laryngoscope 116(11):1956–1961. https://doi.org/10.1097/01.mlg.0000242119.92179.b6
29. Server EA, Alkan Z, Yigit O, Yasak AG (2016) Long-term results of pillar implant procedure. Kulak Burun Bogaz Ihtis Derg 26(5):258–264. https://doi.org/10.5606/kbbihtisas.2016.28924
30. Svirsky MA, Robbins AM, Kirk KI, Pisoni DB, Miyamoto RT (2000) Language development in profoundly deaf children with cochlear implants. Psychol Sci 11(2):153–158. https://doi.org/10.1111/1467-9280.00231
31. House WF (1976) Cochlear implants. Ann Otol Rhinol Laryngol 85(27) (3Pt2):1–93
32. Clark G (2004) Cochlear implants. In: Speech processing in the auditory system. Springer, pp 422–462
33. Wilson BS, Finley CC, Lawson DT, Wolford RD, Eddington DK, Rabinowitz WM (1991) Better speech recognition with cochlear implants. Nature 352(6332):236–238. https://doi.org/10.1038/352236a0
34. Rizzo JF III (2011) Update on retinal prosthetic research: the Boston Retinal Implant Project. J Neuro-ophthalmol 31(2):160–168
35. Chow V, Chow AY (1999) Multi-phasic microphotodiode retinal implant and adaptive imaging retinal stimulation system. Google Patents
36. Zrenner E (2002) Will retinal implants restore vision? Science 295(5557):1022–1025. https://doi.org/10.1126/science.1067996
37. Van de Heyning PH, Badr MS, Baskin JZ, Cramer Bornemann MA, De Backer WA, Dotan Y, Hohenhorst W, Knaack L, Lin HS, Maurer JT, Netzer A, Odland RM, Oliven A, Strohl KP, Vanderveken OM, Verbraecken J, Woodson BT (2012) Implanted upper airway stimulation device for obstructive sleep apnea. Laryngoscope 122(7):1626–1633. https://doi.org/10.1002/lary.23301
38. Strollo PJ Jr, Soose RJ, Maurer JT, de Vries N, Cornelius J, Froymovich O, Hanson RD, Padhya TA, Steward DL, Gillespie MB (2014) Upper-airway stimulation for obstructive sleep apnea. N Engl J Med 370(2):139–149
39. Perlmutter JS, Mink JW (2006) Deep brain stimulation. Annu Rev Neurosci 29:229–257. https://doi.org/10.1146/annurev.neuro.29.051605.112824
40. Deuschl G, Schade-Brittinger C, Krack P, Volkmann J, Schafer H, Botzel K, Daniels C, Deutschlander A, Dillmann U, Eisner W, Gruber D, Hamel W, Herzog J, Hilker R, Klebe S, Kloss M, Koy J, Krause M, Kupsch A, Lorenz D, Lorenzl S, Mehdorn HM, Moringlane JR, Oertel W, Pinsker MO, Reichmann H, Reuss A, Schneider GH, Schnitzler A, Steude U, Sturm V, Timmermann L, Tronnier V, Trottenberg T, Wojtecki L, Wolf E, Poewe W, Voges J (2006) A randomized trial of deep-brain stimulation for Parkinson's disease. N Engl J Med 355(9):896–908. https://doi.org/10.1056/NEJMoa060281
41. Mayberg HS, Lozano AM, Voon V, McNeely HE, Seminowicz D, Hamani C, Schwalb JM, Kennedy SH (2005) Deep brain stimulation for treatment-resistant depression. Neuron 45(5):651–660. https://doi.org/10.1016/j.neuron.2005.02.014
42. Robin S, Sawan M, Harvey J, Abdel-Gawad M, Abdel-Baky T, Elhilai M (1997) A new implantable microstimulator dedicated to selective stimulation of the bladder. In: Proceedings of the 19th annual international conference of the IEEE engineering in medicine and biology society.'magnificent milestones and emerging opportunities in medical engineering'(Cat. No. 97CH36136). IEEE, pp 1792–1795

43. Eisele DW, Schwartz AR, Smith PL (2003) Tongue neuromuscular and direct hypoglossal nerve stimulation for obstructive sleep apnea. Otolaryngol Clin North Am 36(3):501–510. https://doi.org/10.1016/s0030-6665(02)00178-0
44. Keith MW, Peckham PH, Thrope GB, Buckett JR, Stroh KC, Menger V (1988) Functional neuromuscular stimulation neuroprostheses for the tetraplegic hand. Clin Orthop Relat Res 233:25–33
45. Chizek H, Kobetic R, Marsolais E, Abbas J, Donner I, Simon E (1988) Control of functional neuromuscular stimulation systems for standing and locomotion in paraplegics. Proc IEEE 76(9):1155–1165
46. Yarkony GM, Jaeger RJ, Roth E, Kralj AR, Quintern J (1990) Functional neuromuscular stimulation for standing after spinal cord injury. Arch Phys Med Rehabil 71(3):201–206
47. Sheffler LR, Chae J (2007) Neuromuscular electrical stimulation in neurorehabilitation. Muscle Nerve 35(5):562–590. https://doi.org/10.1002/mus.20758
48. Lee RG, van Donkelaar P (1995) Mechanisms underlying functional recovery following stroke. Can J Neurol Sci 22(4):257–263. https://doi.org/10.1017/s0317167100039445
49. Brouillette RT, Thach BT (1980) Control of genioglossus muscle inspiratory activity. J Appl Physiol Respir Environ Exerc Physiol 49(5):801–808. https://doi.org/10.1152/jappl.1980.49.5.801
50. Decker MJ, Haaga J, Arnold JL, Atzberger D, Strohl KP (1993) Functional electrical stimulation and respiration during sleep. J Appl Physiol 75(3):1053–1061. https://doi.org/10.1152/jappl.1993.75.3.1053
51. Fairbanks DW, Fairbanks DN (1993) Neurostimulation for obstructive sleep apnea: investigations. Ear Nose Throat J 72 (1):52–54, 57
52. Schwartz AR, Bennett ML, Smith PL, De Backer W, Hedner J, Boudewyns A, Van de Heyning P, Ejnell H, Hochban W, Knaack L, Podszus T, Penzel T, Peter JH, Goding GS, Erickson DJ, Testerman R, Ottenhoff F, Eisele DW (2001) Therapeutic electrical stimulation of the hypoglossal nerve in obstructive sleep apnea. Arch Otolaryngol Head Neck Surg 127(10):1216–1223. https://doi.org/10.1001/archotol.127.10.1216
53. Kezirian EJ, Boudewyns A, Eisele DW, Schwartz AR, Smith PL, Van de Heyning PH, De Backer WA (2010) Electrical stimulation of the hypoglossal nerve in the treatment of obstructive sleep apnea. Sleep Med Rev 14(5):299–305. https://doi.org/10.1016/j.smrv.2009.10.009
54. Heiser C, Edenharter G, Bas M, Wirth M, Hofauer B (2017) Palatoglossus coupling in selective upper airway stimulation. Laryngoscope 127(10):E378–E383. https://doi.org/10.1002/lary.26487
55. Jordan AS, White DP (2008) Pharyngeal motor control and the pathogenesis of obstructive sleep apnea. Respir Physiol Neurobiol 160(1):1–7. https://doi.org/10.1016/j.resp.2007.07.009
56. Schwartz RS, Salome NN, Ingmundon PT, Rugh JD (1996) Effects of electrical stimulation to the soft palate on snoring and obstructive sleep apnea. J Prosthet Dent 76(3):273–281. https://doi.org/10.1016/s0022-3913(96)90171-7
57. McWhorter AJ, Rowley JA, Eisele DW, Smith PL, Schwartz AR (1999) The effect of tensor veli palatini stimulation on upper airway patency. Arch Otolaryngol Head Neck Surg 125(9):937–940. https://doi.org/10.1001/archotol.125.9.937
58. Gwon TM, Kim C, Shin S, Park JH, Kim JH, Kim SJ (2016) Liquid crystal polymer (LCP)-based neural prosthetic devices. Biomed Eng Lett 6(3):148–163
59. Gwon TM, Min KS, Kim JH, Oh SH, Lee HS, Park MH, Kim SJ (2015) Fabrication and evaluation of an improved polymer-based cochlear electrode array for atraumatic insertion. Biomed Microdevices 17(2):32. https://doi.org/10.1007/s10544-015-9941-1
60. Jeong J, Lee SW, Min KS, Shin S, Jun SB, Kim SJ (2012) Liquid Crystal Polymer(LCP), an attractive substrate for retinal implant. Sensors Mater 24(4):189–203
61. Jeong J, Bae SH, Min KS, Seo J-M, Chung H, Kim SJ (2015) A miniaturized, eye-conformable, and long-term reliable retinal prosthesis using monolithic fabrication of liquid crystal polymer (LCP). IEEE Trans Biomed Eng 62(3):982–989
62. Lee SE, Jun SB, Lee HJ, Kim J, Lee SW, Im C, Shin H-C, Chang JW, Kim SJ (2012) A flexible depth probe using liquid crystal polymer. J IEEE Trans Biomed Eng 59(7):2085–2094

63. Shim S, Park H-Y, Choi GJ, Shin H-C, Kim SJ (2019) A simply fabricated neural probe by laser machining of a thermally laminated gold thin film on transparent cyclic olefin polymer. J ACS Omega 4(2):2590–2595
64. Kuraray (2012) Kuraray LCP Film. https://www.kuraray.eu/fileadmin/presse/publikationen/downloads_k_fair_2013/New_business_development_division/Kuraray_LCP_Film_Brochure.pdf. 20 Feb 2019
65. Van de Graaff WB, Gottfried SB, Mitra J, van Lunteren E, Cherniack NS, Strohl KP (1984) Respiratory function of hyoid muscles and hyoid arch. J Appl Physiol Respir Environ Exerc Physiol 57(1):197–204. https://doi.org/10.1152/jappl.1984.57.1.197
66. Mu L, Sanders I (2010) Human tongue neuroanatomy: Nerve supply and motor endplates. Clin Anat 23(7):777–791. https://doi.org/10.1002/ca.21011
67. Heiser C, Hofauer B, Lozier L, Woodson BT, Stark T (2016) Nerve monitoring-guided selective hypoglossal nerve stimulation in obstructive sleep apnea patients. Laryngoscope 126(12):2852–2858. https://doi.org/10.1002/lary.26026
68. Naples GG, Mortimer JT, Scheiner A, Sweeney JD (1988) A spiral nerve cuff electrode for peripheral nerve stimulation. IEEE Trans Biomed Eng 35(11):905–916. https://doi.org/10.1109/10.8670
69. Haugland MK, Hoffer JA, Sinkjaer T (1994) Skin contact force information in sensory nerve signals recorded by implanted cuff electrodes. IEEE Trans Rehabil Eng 2(1):18–28
70. Rodri FJ, Ceballos D, Schu M, Valero A, Valderrama E, Stieglitz T, Navarro X (2000) Polyimide cuff electrodes for peripheral nerve stimulation 98 (2):105–118
71. Kang X, Liu J-Q, Tian H, Yang B, Nuli Y, Yang C (2015) Self-Closed Parylene Cuff Electrode for Peripheral Nerve Recording. Microelectromech Syst J 24(2):319–332
72. Romero E, Denef JF, Delbeke J, Robert A, Veraart C (2001) Neural morphological effects of long-term implantation of the self-sizing spiral cuff nerve electrode. Med Biol Eng Comput 39(1):90–100. https://doi.org/10.1007/bf02345271
73. Rodri FJ, Ceballos D, Schu M, Valero A, Valderrama E, Stieglitz T, Navarro X (2000) Polyimide cuff electrodes for peripheral nerve stimulation. J Neurosci Methods 98(2):105–118
74. Crampon M, Brailovski V, Sawan M, Trochu F (2001) Nerve cuff electrode with shape memory alloy armature: design and fabrication. Bio-Med Mater Eng 12(4):397–410
75. Crampon MA, Sawan M, Brailovski V, Trochu F (1999) New easy to install nerve cuff electrode using shape memory alloy armature. Artif Organs 23(5):392–395

Chapter 2
Methods

Detailed methods of design, fabrication, and evaluation are provided for the suggested approaches. Overall description of each system is followed by a presentation of the design and fabrication methods of electrodes, the theoretical basis and simulation of circuit design, and the fabrication of the systems. Evaluations are performed not only through bench-top tests, but also in vitro and in vivo tests to demonstrate the feasibility and therapeutic effect of the systems.

2.1 Magnetic Cuff System for HNS

2.1.1 Overview

The proposed system consists of a cuff electrode which provides a novel closing method using magnetic force, and an external pulse generator which generates a train of pulses and delivers it to the electrode. The magnet-embedded nerve cuff electrode (MENCE) has a pair of rare-earth magnets for repeated opening and closing of the cuff. There are a pulse generator ASIC and a timer IC to stimulate the hypoglossal nerve for a second and rest for four seconds.

2.1.2 Magnet-Embedded Nerve Cuff Electrode

2.1.2.1 Electrode Design

Figure 2.1 illustrates an overview of a magnet embedded nerve cuff electrode (MENCE). Three parts consist MENCE: the LCP substrate, gold metal sites, and

J. Seo, *Development of Implantable Electronics as Novel Approaches to Obstructive Sleep Apnea*, Springer Theses, https://doi.org/10.1007/978-981-15-8327-8_2

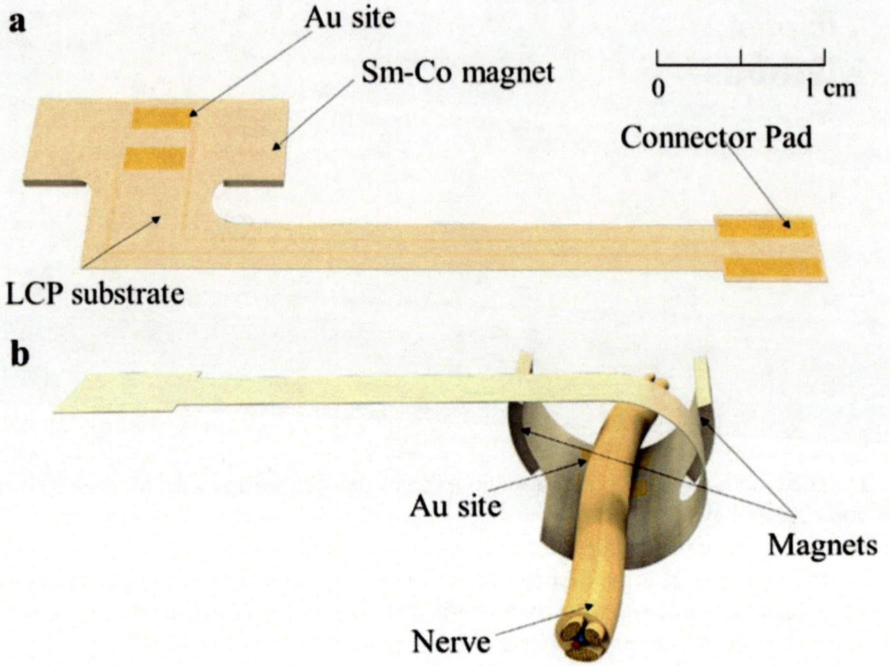

Fig. 2.1 Overview of MENCE: **a** LCP based MENCE in the natural state **b** curled MENCE about to be installed on a nerve

two samarium-cobalt (Sm-Co) magnets. The magnets are fully covered by LCP to achieve hermeticity which is essential for implantation. The substrate of MENCE has 50 μm-thickness, which gives MENCE flexibility to wrap the nerve conformally. There are two Au sites at the cuff and exist two connector pads designed to fit a commercial flat flexible cable (FFC). By combining the pads to an FFC connector, the Au sites at the cuff can be electrically connected to the pulse generator.

2.1.2.2 Fabrication

A brief sequence of the LCP microfabrication process is illustrated in Fig. 2.2. First, silicone elastomer (MED-6233, Nusil Technology, USA) was spin-coated on a bare wafer to attach a bottom layer LCP film with a high melting temperature (HT-LCP; 330 °C, Vecstar CT-Z series, Kuraray, Japan) and was then cured at 100 °C for eight hours. Now the LCP film is compatible with a batch-process using semiconductor microfabrication processes. The LCP film was then cleansed with acetone, methanol, and isopropyl alcohol in sequence for one minute, in each case. The surface of the LCP film was exposed by oxygen plasma (150 W, 100 sccm, 3 min; Plasmalab 80 plus, Oxford Instruments, U.K.) before the metal deposition step for better adhesion. An adhesion layer of 50 nm of Ti and a seed layer of 100 nm of Au were deposited on

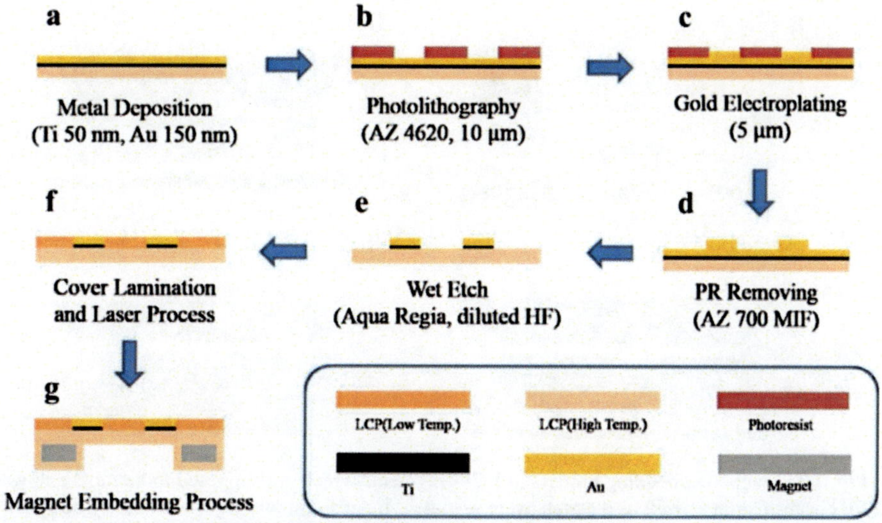

Fig. 2.2 MENCE fabrication process: **a** deposition of a 100 nm Au seed layer on a 500 nm Ti adhesion layer, **b** 10 µm photoresist patterning, **c** 5 µm gold plating, **d** photoresist removal, **e** wet etching of Au seed layer (aqua regia; HCl:HNO3 = 3:1) and Ti wet etching (diluted HF), **f** LCP cover lamination and laser process, and **g** magnet embedding process

the film using an e-gun evaporator (ZZS550-2/D, Maestech). A photoresist (AZ 4620; Clariant, USA) was patterned onto the Au layer with a MA6/BA6 aligner (vacuum contact, 70 s; SUSS MicroTec, Germany) followed by a spin coating process of 40 s at 2000 rpm for 10 µm-thickness. Subsequently, further Au was electroplated to achieve 5 µm thickness. Only on the area where the photoresist was not patterned were electroplated, so the electroplated area becomes fifty times thicker than the non-plated area. After the electroplating process, the photoresist was removed using AZ 700 MIF. Then the Au layers were etched using aqua regia until the 100 nm-thick seed layer completely being etched, thereby only the electroplated area exits after the wet etching. Ti adhesion layer was then etched using diluted hydrofluoric acid. Surface activation and cleaning was conducted by applying oxygen plasma (150 W, 100 sccm, 3 min) as a preparation for the following lamination process. An LCP cover layer film with a low melting temperature (LT-LCP; 280 °C, Vecstar CT-F series, Kuraray, Japan) was prepared. The two LCP films, the bottom layer and the cover layer, were aligned into an aluminum jig and laminated in a heating press (30 min, 285 °C, 4.21 kgf/cm^2; model 4330, Carver, USA). Then the sites which is supposed to be open were ablated using a laser machine (355 nm UV, Samurai system, DPSS, USA). Finally, the electrode was cut along the designed shape.

A process to embed magnets in the cuff, is described from here. A pair of Samarium-Cobalt (Sm-Co) magnets were prepared for embedding magnets, as they have high magnetic flux density and thermal stability. The thermal stability is important because the magnet-embedding process takes place at a high temperature process at 285 °C. Otherwise, the magnets would lose their characteristics after the lamination

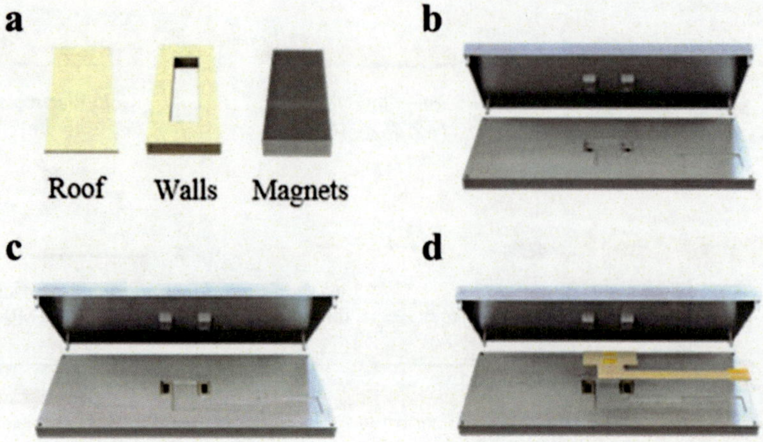

Fig. 2.3 Magnet-embedding process: **a** LCP fragments which are supposed to be roofs and walls for encapsulation as well as magnets were prepared. **b** An aluminum jig was carefully designed to exert pressure onto only the walls in order to prevent the breakage of the magnets. **c** All components were assembled in the jig. **d** Finally, an LCP electrode was joined and laminated

process is carried out. The surface magnetic flux density, the residual flux density, the coercivity, and the intrinsic coercivity of the magnet were 0.140 T, 1.06 T, 0.791, and 2.08 MA/m in each case.

Oxygen plasma (150 W, 100 sccm, 3 min) was treated on a 100 μm-thick HT-LCP film, a 25 μm-thick LT-LCP film and the fabricated LCP electrode to activate their surfaces before the following lamination process. Later, the HT-LCP and the LT-LCP films were cut into either rectangular donut-shaped small pieces or a rectangle-shaped piece, to serve as walls and roofs of the magnet encapsulation. The LT-LCP pieces were used as adhesives to glue HT-LCP films, as the LT-LCP melts at 285 °C while HT-LCP does not. The electrode and the cut pieces were placed together in a specially designed aluminum jig and laminated by a heating press (30 min, 285 °C 15.65 kgf/cm^2). The overall magnet-embedding process is depicted in Fig. 2.3 for a better understanding of the described process.

2.1.3 *External Pulse Generator*

2.1.3.1 Circuit Description

An ASIC (Samsung CMOS 0.18 μm) in the reference was used for generating a biphasic current pulse. As the ASIC inputs parallel bits of parameters, amplitude, duration, and pulse rate, mechanical switches were used for setting the parameters. A commercial timer IC (LM555CM, Texas Instruments, U.S.) was used to control durations of 'ON' and 'OFF' states of the stimulation, one second and four seconds, respectively.

2.1.3.2 Fabrication

The pulse generator was fabricated on a printed circuit board having a dimension of 100 mm × 50 mm including a battery holder of 50 mm × 50 mm. The ASIC was wire-bonded on the board while all other components were soldered. Two male connectors were connected to the outputs of the ASIC via DC-blocking capacitors which was used for safety of the stimulation.

2.1.4 Evaluations

2.1.4.1 Electrochemical Measurements in Vitro

The electrochemical impedance spectrums (EISs) and cyclic voltammetries (CVs) were measured (n = 10) using the Solatron 1260 and 1287 systems (Solatron, U.K.) in a three-electrode cell using an Ag|AgCl reference electrode and a mesh-type Pt counter electrode in PBS solution (Gibco, Thermo Fisher Scientific, U.S). Frequency was swept from 0.1 Hz to 10 kHz by applying a sinusoidal of 10 mV amplitude with a zero offset during the measurement. The CV curve was recorded in the range of −0.6 – 0.8 V at a scan rate of 1 mV/s at which the CVs of the electrodes can be measured in the steady state [1]. The cathodic charge storage capacity (CSC_c) was calculated by integrating the cathodic area of the plotted CV curve using the CorrView software package. The setup of the in vitro evaluation is depicted in Fig. 2.4.

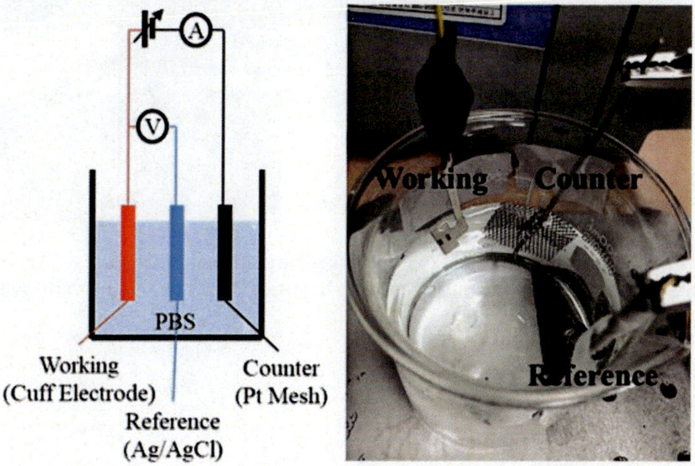

Fig. 2.4 Illustration (left) and picture (right) of in vitro electrochemical measurement setup of magnetic cuff using three-electrode cell

2.1.4.2 Animal Testing in Vivo

All the following experimental procedures were approved by the International Animal Care and Use Committee in Seoul National University Bundang Hospital. Seven-month-old New Zealand White male rabbits weighing 2.1–2.6 kg were used in this study. All rabbits were acclimated in the new environment for the first week and used for further experiments. Sedation was induced by intramuscular injection of 0.3 mL/kg of Zoletil (tiletamine hydrochloride plus zolazepam hydrochloride; Virbac) and 0.25 mL/kg of Rompun (xylazine; Bayer AG) at the same time. After sleep was induced, a midline neck incision was made around hyoid bone. The main trunk of the hypoglossal nerve was identified at the lateral side of the hypoglossus muscle over the external carotid artery. The identified nerve was dissected to place MENCE which was percutaneously connected to an external pulse generator through an FFC connector. The pulse generator was custom-made to support varying amplitudes, duty cycles, frequencies and pulse rates of the output stimulation current. Pulse amplitude, rate, and duration of the current stimuli of were 0.106 mA, 0.635 kHz and 0.375 ms, respectively. The stimuli were applied to the nerve for 1.15 s with a rest of 3.93 s. Status of the upper airway of the rabbit was monitored using C-arm fluoroscopy (Philips, Netherlands) during the stimulation. The experimental setup is depicted in Fig. 2.5.

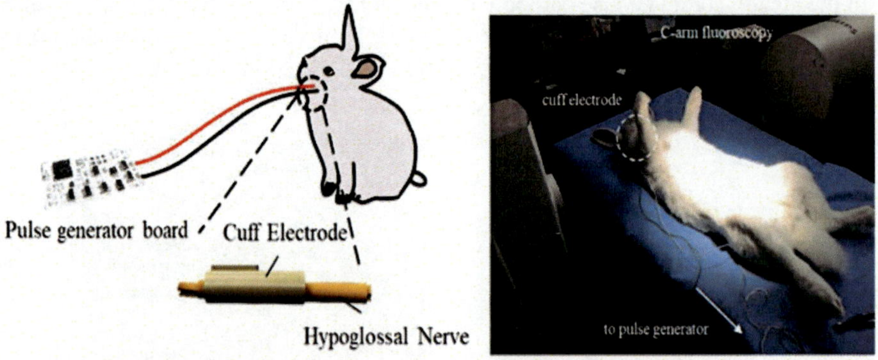

Fig. 2.5 Illustration of the in vivo experimental setup (left) and its picture (right). Installed cuff was percutaneously connected to pulse generator board and behavior of upper airway was recorded via C-arm fluoroscopy

2.2 Palatal Implant System for SPS

2.2.1 Pilot Study in Vivo

Two electrodes serving as a stimulation electrode and a reference electrode, were prepared for the experiment (needle-type, 20°, A-M Systems Inc., U.S.). Each electrode was made of stainless steel with a diameter of 0.254 mm and exposed at the end of the tip by 3 mm. A custom pulse generator was fabricated to deliver current stimuli to the electrode. A current generation ASIC with parameter-controlling interfaces were surface-mounted on the PCB of the pulse generator [2].

A seven-month-old New Zealand white male rabbit was prepared for the experiment. The rabbit was acclimated to the new environment for the first week. Sedation was induced by intramuscular injection of 0.3 ml/kg of Zoletil (tiletamine hydrochloride, zolazepam hydrochloride; Virbac) and 0.25 ml/kg of Rompun (xylazine; Bayer AG) at the same time. After sleep was induced, the prepared needle-type electrodes were inserted into the soft palate of the rabbit. Then, current stimuli generated from a custom-made pulse generator, was applied to the soft palate through the electrodes. While the compliance voltage level was maintained to 9 V and the duration set at 636.3 μs, the frequency was swept from 20 Hz to 200 Hz to find the most effective frequency range. Next, to find the threshold level, the current level was gradually decreased. During the stimulation, the upper airway of the rabbit was observed with C-arm fluoroscopy (Philips, Netherlands). The described experimental setup is shown in Fig. 2.6.

To figure out a long-term reliability of the implant under in vivo condition, a test inserting a mock-up device was performed. The fabrication process of the mock-up is following. First, high-temperature (HT-LCP; 330 °C, Vecstar CT-Z series, Kuraray, Japan) and low-temperature (LT-LCP; 280 °C, Vecstar CT-F series, Kuraray, Japan)

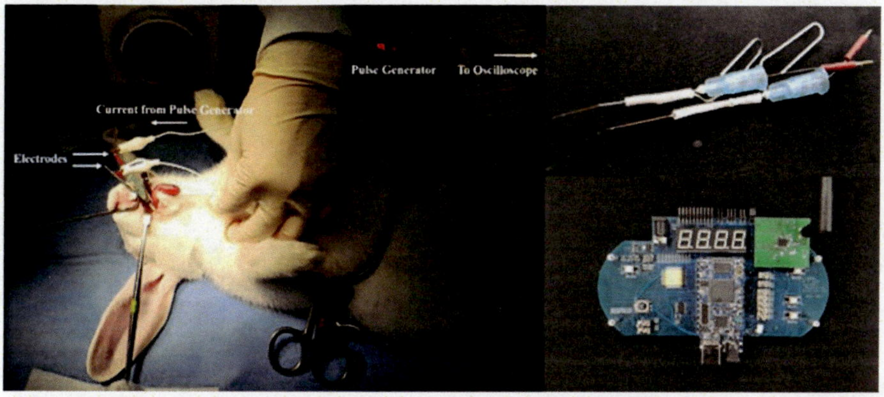

Fig. 2.6 In vivo test setup of the pilot study (left) using commercial electrodes (right-top) and custom-made pulse generator (right-bottom)

LCP films with 100 μm-thickness were prepared. The films were cut along a desired shape using a laser machine (355 nm UV; Samurai system, DPSS, U.S.). The films were then cleansed with acetone, methanol, and isopropyl alcohol in sequence each for a minute. Next, the films were exposed to oxygen plasma (150 W, 100 sccm, 3 min; Plasmalab 80 plus, Oxford Instruments, U.K) as a preparation step to the following lamination process. Subsequently, the HT-LCP and the LT-LCP pieces were stacked in a custom-designed aluminum jig which was coated with Teflon of 0.1 mm thickness to prevent the LCP pieces from sticking to the jig during the lamination process. The pieces were laminated under pressure and temperature of 3 MPa and 285 °C each, for 30 min. The laminated mock-up was then autoclaved at 121 °C for 20 min to remove any possibility of infection during implantation (Unisteri, BMT Medical Technology, CZE). The fabrication process is illustrated in Fig. 2.7.

A New Zealand white rabbit was prepared and induced to the sleep following the same procedure stated in Sect.2.1.4.2. An incision was made along the lateral axis of the hard palate, to and then the mock-up was inserted toward the soft palate from the incision point. After the insertion, the wound was closed and sutured. Antibiotics with 50 mg/ml (enrofloxacin; Bayer AG) were injected to the rabbit every day after the surgery to prevent further inflammations. All the experimental procedures were approved by the International Animal Care and Use Committee in Seoul National University Bundang Hospital.

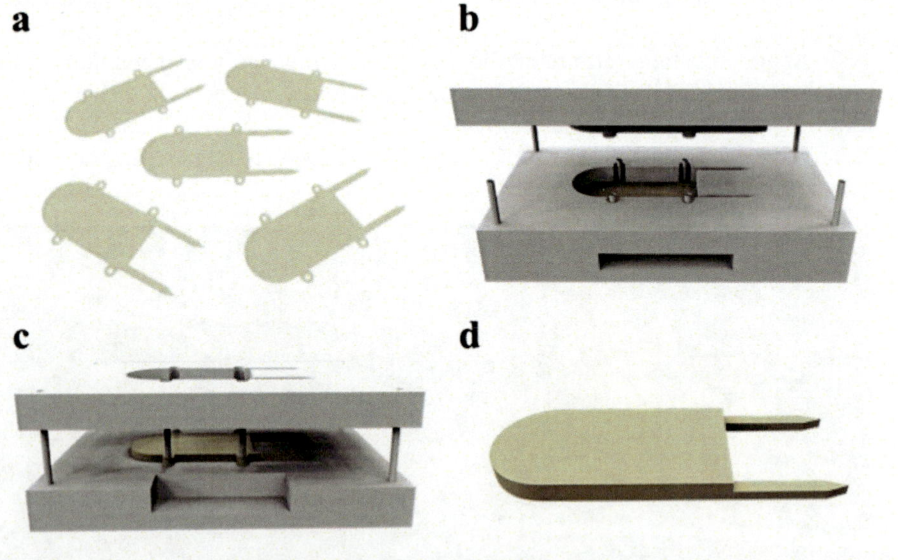

Fig. 2.7 Fabrication process of palatal implant: **a** preparation of LCP pieces, **b** lamination, **c** lift-up of the jig sidewall, and **d** completion of the fabrication with laser-machining

2.2.2 Overview

Figure 2.8 provides an overview of the proposed palatal implant system. The system consists of two major parts, namely the implant and the intra-oral device (IOD). The implant has a multi-layered structure consisting of a coil, electrodes, and a circuit. Two electrodes exist on each shank upward and downward to generate a bipolar current around the tip. The implant receives power and data via an inductive link from the IOD through the hard palate skin and delivers generated current stimulation pulses to the electrode interfacing the soft palate.

The IOD is a device in the shape of an oral appliance, which can be equipped in the mouth. It contains a transmitter coil, PCB, battery, and wireless charging coil, all of which are fully packaged by silicone elastomer (MED-6233, NuSil Technology Inc., U.S.). In the circuit, there is a commercial ZigBee transceiver IC (CC2530, Texas Instrument, U.S.), which receives a command signal transmitted from the controller. The received command signal is then modulated and amplified by a class-E amplifier and delivered to the implant via an inductive link. The circuit in the IOD is fully-powered by an embedded rechargeable battery with a 1.11-Wh capacity.

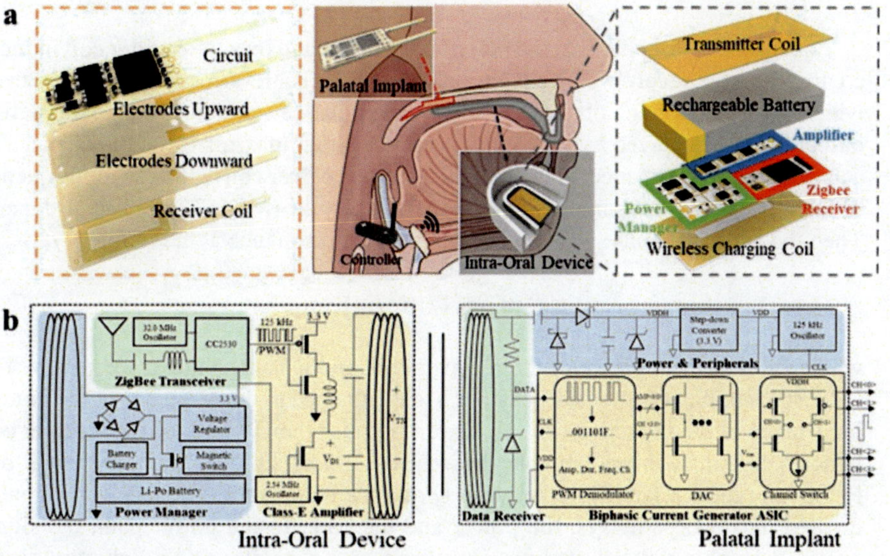

Fig. 2.8 Overview of the proposed palatal implant system. **a** Overview of the palatal implant system. The palatal implant has a pair of electrodes at the end of each shank to stimulate the soft palate. The intra-oral device, designed to be equipped in the oral cavity, transmits RF signals to the implant to operate it. The detailed structure of each device is illustrated at the left and the right. **b** Circuit diagrams of the system. The intra-oral device modulates and amplifies a data signal received via ZigBee. Then the implant utilizes the received signal as both power and data to operate the ASIC that supplies biphasic current pulses to the electrodes

An auxiliary device, the wireless controller, is a portable device that provides a user-interface for setting stimulation parameters, such as duration, frequency, and amplitude. The determined parameters, which serve as a command signal to the ASIC in the implant, are transmitted to the IOD via ZigBee.

2.2.3 Palatal Implant

2.2.3.1 Circuit Description

The circuit of the implant consists of an RF receiver that receives a modulated signal and a stimulation current pulse generator ASIC with its peripheral components. First, the RF receiver comprises a planar coil and a capacitor that produce parallel resonance whose impedance is maximized at the resonance frequency. According to Ampere's circuital law, the induced current density of a coil has the following relationship with the magnetic flux intensity H:

$$\nabla \times H = J. \tag{2.1}$$

Equation (2.1) implies that the magnetic field created by a transmitter coil induces a current on the receiver coil via an inductive link. Therefore, a high voltage level, which is necessary for obtaining enough compliance voltage for the stimulation output, can be achieved by the parallel resonator that maximizes its impedance. To achieve parallel resonance, the inductance of the receiver coil at the carrier frequency of 2.54 MHz was measured using a network analyzer (MS4630B, Anritsu, Japan). Then the capacitor value was selected to produce resonance at the frequency:

$$C = 1/\left(w^2 L_R\right), \tag{2.2}$$

where L_R is the inductance of the receiver coil, and w is the angular frequency. The LC resonance can be affected by either a load attached to the resonator or the mutual inductance of the inductive link. However, because the Q-factor of the receiver coil is normally small in implantable devices, the efficiency does not vary much due to the reciprocal relationship between Q and the bandwidth. The received signal is divided into two pathways, the power and the path. In the power path, the signal passes through a voltage clamper consisting of a capacitor and a Schottky diode. The voltage clamper brings the bottom peak of the voltage to the reference level, literally 'clamping' the voltage to the reference. The highest peak voltage becomes approximately twice, specifically $2V_p - V_f$, where V_p is the peak voltage and V_f is the forward bias of the diode. The clamped voltage is then rectified by the half-wave rectifier, creating VDDH the compliance level of the stimulation output. As the voltage is clamped to positive, there is little power loss except the dissipated power of the diode. A Zener diode with a Zener-voltage of 20 V is added to the rectifier for safety, so as not to exceed the maximum acceptance voltage level of the ASIC.

VDDH is further fed to a step-down converter (NCP715, ON Semiconductor, U.S.). The step-down converter can receive an input voltage up to 20 V and output VDD the constant voltage level of 3.3 V. The generated VDD is supplied to both the ASIC and the oscillator (LTC6906, Analog Devices, U.S.). The oscillator is programmed by a resistor to input a 2.54-MHz pulse rate. An additional bypass capacitor is added parallel to the resistor to eliminate the AC interference caused by the magnetic field of the receiver coil. The output pulse is used as a clock of the digital part of the ASIC. On the other hand, a simple voltage clipper is implemented in the data path. The signal is clipped by the Zener diode to limit the voltage level of the DATA to the VDD to be fed into the digital part of the ASIC. The value of the resistor in the clipper should be carefully selected. If the resistance becomes too small, the voltage across the resonator decreases as the decreased resistance determines the equivalent load of the circuit. If the resistance becomes too large, the clipped voltage may decrease under the threshold of the ASIC. The resistance value was set to 1 kΩ, experimentally.

The ASIC inputs VDDH, VDD, and DATA to generate a biphasic stimulation current pulse. The ASIC extracts frequency, duration, and level information of the output pulse from DATA. According to the information, the ASIC generates an output pulse having a compliance level of VDDH. Figure 2.9 illustrates the description.

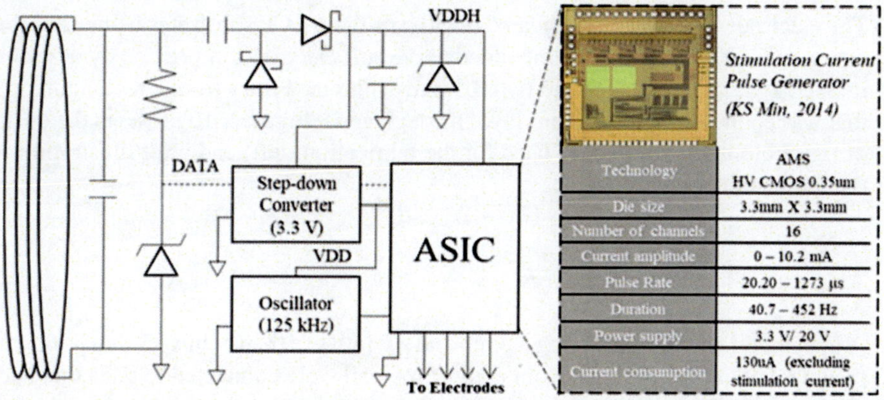

Fig. 2.9 Circuit diagram of the palatal implant. The RF receiver comprised of LC parallel resonance gains high levels of the voltage. The receiver rectifies and clips to use the signal as power and data for current generator ASIC. The ASIC is capable of generate biphasic current pulses with variable amplitude, pulse rate, and duration, as arranged in the rightmost table

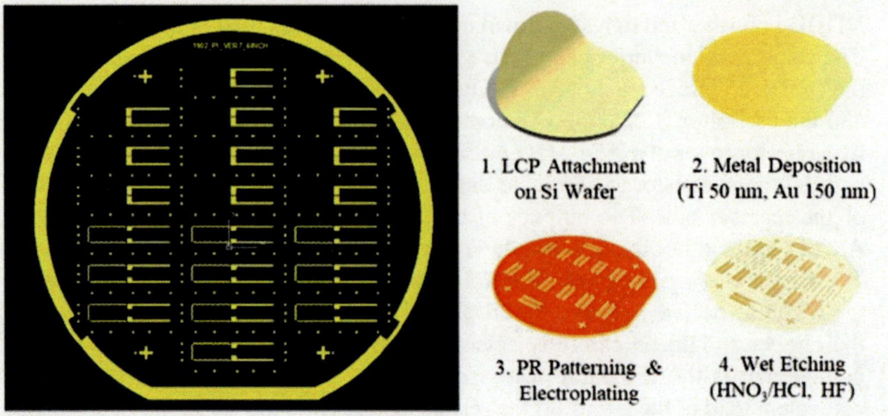

Fig. 2.10 Photolithography mask of the electrode layer (left) and brief illustrations of the fabrication process of the palatal implant

2.2.3.2 Fabrication

Electrode Layers

The electrode layers were fabricated on 50 μm-thick LCP substrates using the fabrication process of the magnet embedded nerve cuff electrode, as previously described in Sect. 2.1.2.2. Each layer had two 0.5 mm-width lead lines from a rectangular pad that was connected to the circuit layer through a via. Figure 2.10 presents the design of the photolithography mask used for the fabrication (left) and brief illustrations of the fabrication method (right).

Antenna Coil and Circuit Layers

A planar coil was drawn having a pitch and a width of 100 μm in a 13 mm × 6.5 mm rectangle with twenty turns. Both ends of the coil were connected to pads that had a diameter of 1.0 mm with a 0.2-mm drill. Two additional pads were patterned on the right side of the coil as stoppers in a laser process, which will be described later.

The circuit described in 2.2.3.1 was drawn in a layout reflecting all physical dimensions including pads and packages. The commercial chip packages and passive components were carefully selected to minimize their footprints on the limited board size. The circuit had six via pads, two for connecting the coil and four for connecting the electrodes. The design of the antenna coil and the circuit are shown in Fig. 2.11 (left).

A conventional PCB fabrication process was used to fabricate the antenna coil layer and a circuit layer. A double-sided, copper-clad LCP sheet was prepared (Ultralam 3850, Rogers Corporation, US). On the 50-μm-thick LCP substrate, a 17.8-μm-thick copper sheet was clad on each side. The metals on the both sides

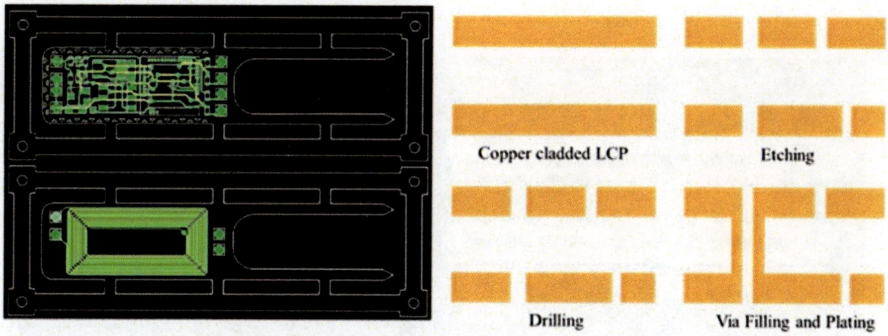

Fig. 2.11 Layouts of circuit and coil layers of the palatal implant (left) and its fabrication methods (right)

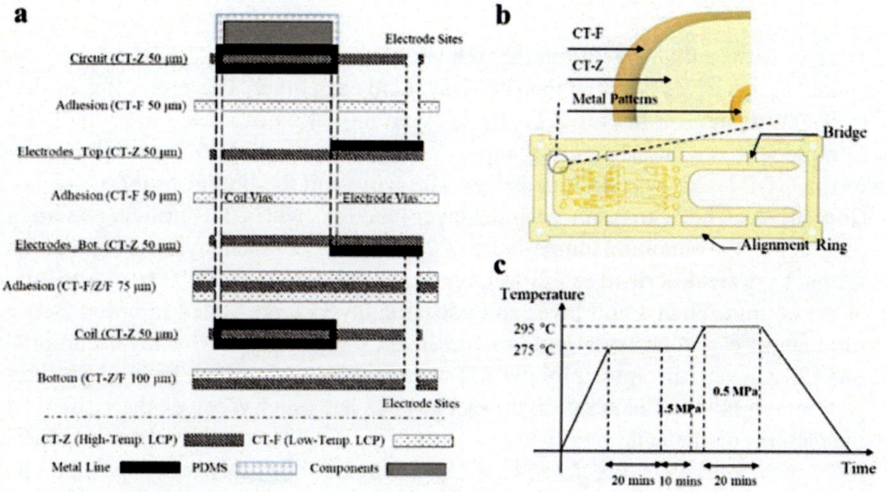

Fig. 2.12 **a** Lamination profile of the palatal implant fabrication. Multiple LCP sheets including functional layers as well as adhesion layers were aligned and laminated. **b** High temperature (CT-Z) and low temperature (CT-F) LCPs were stacked in turn. **c** Plot of temperature and pressure condition for the lamination

were etched and patterned using a mask. Then vias were drilled by a UV laser and filled with copper. Finally, an additional 17.8-μm-thick Au layer was plated on the copper. The fabrication process is depicted in Fig. 2.11 (right).

Multi-layer Lamination

The implant was fabricated by laminating multiple LCP layers, including a coil, a circuit, electrodes, and insulation layers (Fig. 2.12a). Because the laminating temper-

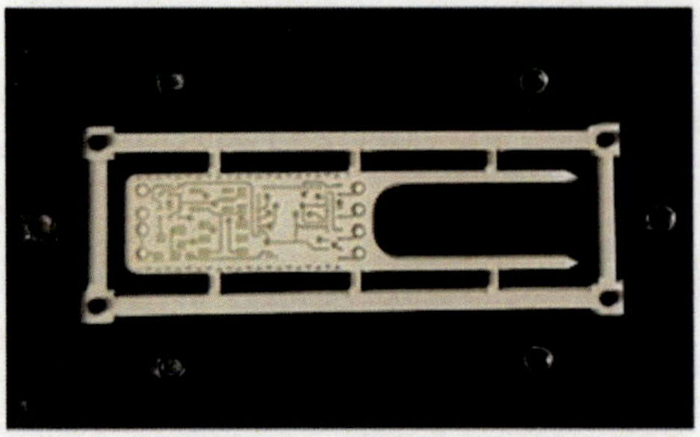

Fig. 2.13 LCP sheets aligned and stacked into the jig (left) and illustrated parts of the jig (right)

ature is between the melting temperatures of the CT-Z and CT-F, the CT-F layers served as adhesives bonding the CT-Z layers to each other. Therefore, it is desirable to pattern metals on the CT-Z layers to avoid migration of lines during the lamination process, not on CT-F. CT-F and CT-Z layers were stacked in turn, but the size of the CT-F layers was designed to be 100 μm from the border of the CT-Z layers (Fig. 2.12b). This is to cover the inter-layer junctions with CT-F, utilizing its melting property on the operation temperature of 290 °C. The detailed layer configurations of the package are described as follows. First, an insulation layer of 100 μm was formed at the bottom. Then a coil layer and adhesion layers were added in order. Between the bottom electrode layer and the coil layer, there was a buffer layer comprising one CT-Z layer surrounded by two CT-F layers. The role of the buffer layer was to prevent migration of patterns on the electrode layer, which receives sheer stress from the patterns on the coil layer. The electrode layer was then deposited on the adhesion layer with patterns facing downward. Another electrode layer was deposited on the adhesion layer facing upward, and finally a circuit layer was added at the top.

After all the layers were stacked, heat and pressure were applied to the layers to fabricate them into one body. The lamination conditions of pressure and temperature are plotted in Fig. 2.12c. First, the temperature was increased from room temperature without pressure. When the temperature reached 275 °C, the sample remained untreated for 20 min to make sure that the sample reached the temperature. Then a pressure of 1.5 MPa was applied to the sample for 10 min. As both CT-F and CT-Z were crystalized at that temperature, they bonded when pressure was applied. The temperature was further increased to 295 °C where CT-F totally melts. The pressure of 0.5 MPa was applied to the sample to fill the spaces between the layers for 20 min. Finally, the sample was cooled down slowly to room temperature.

Figure 2.13 presents the aluminum lamination jig used for lamination. The jig was coated by Teflon to prevent adherence of the LCP to the aluminum after lamination. The jig consisted of three parts; the bottom part for loading the LCP layers, the middle part for limiting spread of the melted CT-F, and the top part for applying force to the LCP sample. First, the LCP samples were loaded into the bottom part of the jig. In the bottom part, there were four stainless-steel poles to fix the position. After the LCP samples were loaded into the jig, the middle part was joined to the bottom part with screws. If the parts are not screwed, melted CT-F may flow through the small gap between the parts. Finally, the top part was assembled to the middle-bottom part along the assembly poles. Ceramic sponges covered by aluminum foil were placed on the top and under the bottom part to apply the force evenly to the sample during lamination.

Post-lamination Process

After lamination was completed, vias were created to connect the layers to each other. A laser machine (355 nm UV, Samurai system, DPSS, USA) was used for the via creation parameters having a laser power of 400 mW, a mark speed of 500 mm/s, a frequency of 20 kHz, and a pulse width of 20 μs. Holes with a diameter of 700 μm were drilled from the circuit to the coil layer and to the electrode layers until the pads were visible through a microscope. Lines with 180 μm-spacing were drawn using a CAD program (AutoCAD, Autodesk, U.S.) to generate the ablation pattern. The electrode sites were also opened using the parameters stated above. A rectangular area of 2 mm $\times$ 0.5 mm was ablated for each electrode site. After the laser ablation process was done, the sample was cleaned using oxygen plasma (150 W, 100 sccm, 3 min; Plasmalab 80 plus, Oxford Instruments, U.K.). Then iridium oxide was electroplated from an aqueous solution of 4 mM $IrCl_4$ in a supporting electrolyte of 40 mM oxalic acid and 340 mM K_2CO_3, following the method our group previously published [3]. Fifty instances of triangular waveforms ranging from 0 to 0.55 V (vs. Ag/AgCl reference) at a 50-mV/s sweep rate were followed by 1800 rectangular pulses at 0.55 V (1 Hz frequency, 0.5 s duration).

The constructed vias were filled with silver epoxy (H20E, Epoxy Technology, Inc., U.S.) to electrically connect the layers. Thus, layers of coil-circuit-electrode could be electrically connected through the filled vias. Then the silver epoxy was cured for 12 h at 50 °C. After the curing was done, measurements of resistance and inductance were conducted to assess the sample. Electronic components having minimized footprints were selected to reduce the layout dimensions of the sample. The components were soldered at 300 °C, where high-temperature LCP (CT-Z) substrates do not melt. Finally, the package was encapsulated by silicone elastomer (MED-6233, NuSil Technology Inc. U.S.) except the electrode shanks. Two parts of silicone (MED-6233A, MED-6233B) were mixed with a 1:1 ratio using an electric mixer. The mixed silicone was loaded into a centrifuge and then loaded into a vacuum chamber to extract bubbles from it. An aluminum mold was designed and manufactured for the packaging. The silicone elastomer was injected through the feed at the top to fill the cavity in the mold. Two air outlets were drilled to allow air to escape so that no air would be trapped in the elastomer.

Table 2.1 Digital protocol of the pulse generator ASIC. Bit '17' and '0' means the most significant bit (MSB) and the least significant bit (LSB), respectively. Prefix 'R', 'I', 'D', 'C', 'A', 'S' means pulse rate, index, duration, current source, amplitude, and site, respectively. 'F' and 'P' stands for announcing start/end of the frame and parity check

Bits	17	16	15	14	13	12	11	10	9	8	7	6	5	4	3	2	1	0
SET1	F	0	0	0	x	x	R0	R1	R2	R3	R4	R5	x	x	x	x	P	F
SET2	F	0	0	1	I0	I1	D0	D1	D2	D3	D4	D5	x	X	x	x	P	F
SET3	F	0	1	0	C1	C2	A0	A1	A2	A3	A4	A5	A6	A7	A8	A6	P	F
SET4	F	0	1	1	C1	C2	S0	S1	S2	S3	I0	I1	B	ON	x	x	P	F
STM1	F	1	0	0	x	x	x	x	x	x	x	x	x	x	x	x	P	F
STM2	F	1	0	1	x	x	x	x	x	x	x	x	x	x	x	x	P	F

2.2.4 *Intra-Oral Device*

2.2.4.1 Circuit Description

In the intra-oral device (IOD), a wireless charging coil, rechargeable battery, and transmitter coil were connected to a printed circuit board (PCB). As the whole parts of the IOD were packaged by silicone elastomer for oral-equipment, a wireless charging technique was adopted for the rechargeable Li-Po battery. A command signal, received via ZigBee from the external controller, was ASK modulated and amplified by a class-E amplifier. Then the modulated signal was transmitted to the receiver coil in the palatal implant via an inductive link.

ZigBee Receiver

A commercial ZigBee compliant RF transceiver chip and antenna were used to implement the ZigBee receiver (CC2530, Texas Instrument, USA; 2450AT42A100E, Johanson Technology Inc., USA). The transceiver includes an 8051 CPU core and a 256 Kb flash programmable memory. The transceiver was programmed to generate PWM based on the message signal from an external controller. The message signal contained information about parameters of the stimulation current pulse, including the duration, frequency, amplitude, and output channels of the pulse. Table 2.1 shows the protocol of the message signal. There were 'F' bits located at both ends of the word to inform the ASIC of either the start or the end of the word. Also, 'P' bits were located at the second least-significant bit to perform an even parity check. Each word consisted of 18 bits with six modes, namely, four modes for settings and two for stimulations. The parameters of the stimulation pulse could be set by inputting SET 1–4 and the stimulation could be started or stopped with STM 1–2. The array of the words generated by the ZigBee transceiver was fed into the ASK modulator to modulate the message on the carrier wave.

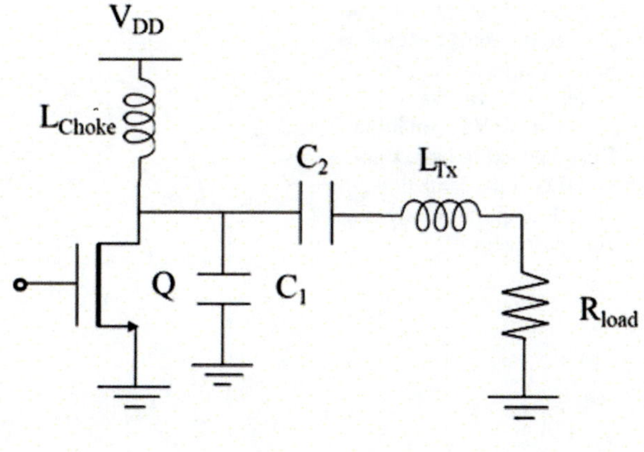

Fig. 2.14 Schematic of basic class-E amplifier. L_{choke} serve as DC pass and C_1, C_2, LTX comprises a resonance at both ON and OFF state of the transistor Q

Class-E Amplifier

Class-E amplifiers are a popular amplifier class for RF applications owing to their high efficiency, ideally 100%. A class-E amplifier operates in a switched-mode where the transistor behaves as a switch with rail-to-rail pulse gate drivers. It differs from other switched mode amplifiers, such as a class-D amplifier, in that the class-E amplifier utilizes a shunt capacitor, including the parasitic capacitor of Q, as a resonant agent. In the other classes, the shunt capacitor in Q is considered as a 'bad' component that needs to be removed because it induces power loss by preventing a sharp transition between ON and OFF. However, the class-E amplifier overcomes the problem by utilizing a shunt capacitor as a resonance agent. The capacitors C_1 and C_2-in Fig. 2.14 create resonance with L_{Tx}, the transmitter coil, while L_{choke} feeds a constant current to Q. There are two resonance frequencies depending on the state of Q. When Q is ON, C_1 is virtually shunt out as the drain and the source of Q are connected through the on-resistance which is normally under a few ohms. Therefore, only C_2 and L_{Tx} participate in the on-state resonant frequency:

$$f_{on} = \frac{1}{2\pi C_2 L_{Tx}} \tag{2.3}$$

On the other hand, when Q is OFF, C_1, C_2, and L_{Tx} participate in the off-state resonant frequency as follows:

$$f_{off} = \frac{1}{2\pi (C1||C_2) L_{Tx}}, \tag{2.4}$$

while L_{choke} is assumed to be large enough that AC signals do not pass through. For the current to have a sinusoidal waveform, the switching frequency f_{sw} should be between (2.3) and (2.4), that is,

$$f_{on} < f_{sw} < f_{off}. \tag{2.5}$$

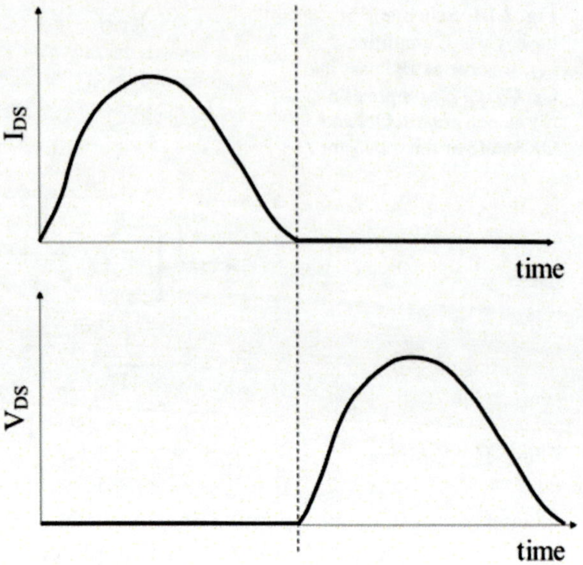

Fig. 2.15 Current (top) and voltage (bottom) waveforms at the drain-source nodes of Q under zero-voltage switching (ZVS) condition. Designed not to have VDS and IDS at the same time, the class-E amplifier ideally has zero power loss

It is desirable for f_{on} and f_{off} to be within 10% of f_{sw} meeting an arithmetic mean in practice. From Eqs. (2.3–2.5), the initial values of C_1 and C_2 can be defined.

For the amplifier to be efficient, the switching loss of Q should be minimized. Switching loss usually occurs when the state of the transistor starts to switch, when I_D and V_{DS} are likely to exist at the same time. Thus, it is important to meet a zero-voltage switching (ZVS) condition, as shown in Fig. 2.15. The key point of ZVS is to delay the rise of I_D until V_{DS} reaches zero and to delay the rise of V_{DS} until I_D reaches zero, allowing both transitions only at $V_{DS} = 0$. Thus, the power dissipation at the transistor can become ideally zero, neglecting its ON-resistance, R_{on}. The ZVS of the amplifier depends on the value of R_{load} including equivalent series resistances (ESR) of the capacitors, the resistance of L_{Tx}, and the reflected resistance from the counterpart coil. Therefore, further fine adjustment of C_1 and C_2 is further required in addition to condition (2.5). The detailed tuning method is well described in a previous paper, providing a guideline to tune C_1 and C_2 according to the measured voltage waveforms of the transistor [4]. The tuning procedure is quite tricky though, as both the condition (2.5) and ZVS should be met simultaneously through iterations. It is recommended to use a transistor with a low output capacitance, C_{oss}, compared to C_1. Because C_{oss} affects the resonance when the transistor is OFF, it may results in a discrepancy between the calculated value and the actual value of the output frequency if C_{oss} is too high. However, transistors with low C_{oss} normally have larger R_{on}, which dissipates the power in terms of heat in every half period. Therefore, it is important to select an appropriate transistor considering the trade-off.

ASK Modulation

Amplitude-shift keying (ASK) is a kind of amplitude modulation that represents digital data as amplitude variations of a carrier wave. An advantage of ASK is simplicity in implementation, requiring only a pair of MOSFETs comprising an inverter added to the power supply. The power supplied to the amplifier is synchronized to the control signal driving the gate; consequently, the envelope of the carrier wave can be modulated by the signal. Although ASK has poor accuracy compared to other modulation methods, such as the frequency-shift keying (FSK) and phase-shift keying (PSK), its simplicity provides flexibility and feasibility in designing a low-power miniaturized circuit.

Power Management

The intra-oral device (IOD) contained a rechargeable battery because the IOD delivers power to the implant. The whole IOD was encapsulated by silicone elastomer to prevent penetration of saliva. Therefore, power management, including charging and power switching should be enabled in non-contact methods. Strategies of inductive link powering and magnetic switching were used for this purpose.

A wireless charger unit based on a class-E amplifier was designed and fabricated to transmit a 2.54-MHz electromagnetic signal to the IOD. A heat sink was mounted on the ground pad of the board using a silver epoxy (H20E, Epoxy Technology Inc., U.S.) to prevent overheating. A commercial transmitter coil (760308104113, Wurth Electronics Inc., Germany) was used for the transmitter antenna. For higher charging efficiency, the optimal geometric values were selected from [5].

The transmitted electromagnetic signal was received by the receiver coil (AWCCA-28R15H08-C01-B, Abracon LLC, U.S.) in the IOD. The coil produced parallel resonance with a capacitor having the value of $1/(w^2L)$, where w is an angular velocity at 2.54 MHz. The received RF signal was rectified by a full-wave rectifier followed by a Zener diode with a Zener voltage of 6 V. The diode was added to avoid an overvoltage to protect a battery charger IC (BQ21040, Texas Instruments Inc., U.S.). The battery charger delivered the rectified power to the lithium polymer battery; thus, the battery could be recharged.

A magnetic switch was implemented to control the power. The switch comprised a magnetic sensor and a D flip-flop to behave as a toggle switch. The sensor printed a digitized output after comparing the detected magnetic flux and its threshold level. Then the output was fed to the clock pin of the D flip-flop whose output was connected to the input 'D' to implement the toggle behavior. In this principle, the power of the IOD could be switched by a permanent magnet in a non-contact method.

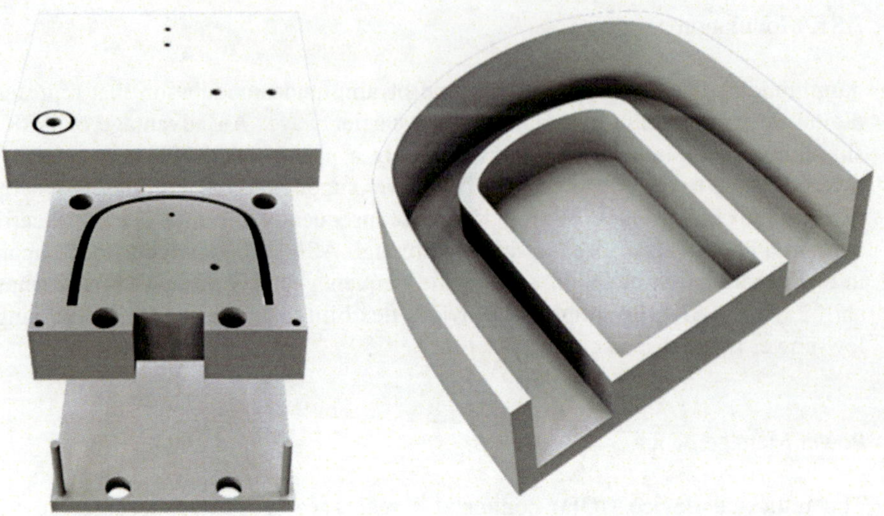

Fig. 2.16 Design of aluminum mold depicting separated parts (left) and illustration of silicone frame of IOD. Electronics as well as battery is to be placed in the empty space in the middle

2.2.4.2 Fabrication

The described circuit was fabricated on a printed circuit board and all the components comprising the ZigBee IC, class-E amplifiers, transmitter coil, and power management circuits were soldered on the board manually. Finally, a 1.11-Wh lithium-polymer battery (TW 801630, Taiwoo, China) was soldered to the board.

The fabricated IOD should be encapsulated as it is intended to be applied in the mouth. The design of the mold and its output are presented in Fig. 2.16. Medical-grade silicone elastomer (MED-6233, NuSil Technology Inc. U.S.) was used to fabricate the shape following the method described in 2.2.3.2.4. In the empty space of the fabricated frame, the electronic components of the IOD and a battery were inserted and further encapsulated by the silicone elastomer.

2.2.4.3 Inductive Link

The palate does not have enough spatial capacity for bulky devices to be implanted. Therefore, it is necessary to eliminate a battery from the implant for miniaturization. Furthermore, conventional wireless communications using the ISM band of 2.4–2.485 GHz, requires IC chips and their peripheral components to have relatively large footprints compared to the available dimension of the implant. Because the first priority is to meet the dimensional criteria, both battery power and conventional communication technology were not used in the implant. Instead, the implant was designed to receive power and data simultaneously from an RF receive coil via an inductive link.

2.2.4.4 Design of Coil Antennas

Coils are widely used as antennas for wireless power transmission via an inductive link. The transmission design for our application required special consideration in that it is used in a bio-medium and data is transmitted simultaneously with power.

Tissue absorption of EM waves tends to increase as the wave frequency increases. However, the tendency is not correct in every frequency range. This is because low-frequency waves reflect on the kin-air interface, while a high-frequency wave is absorbed by tissues. Indeed, Vaillancourt demonstrated that the minimum power loss can be achieved near a few mega-hertz from his research. From the power-loss graph in the literature, the RF frequency of 2.54 MHz was selected for our application [6].

Implantable devices normally have strict dimensional restrictions depending on the implantation site. From our pilot study, the dimensions of the receiver antenna were chosen to be 13 mm × 6.5 mm. The receiver antenna was designed to be a rectangular planar-type coil patterned on both sides of an LCP sheet, when the number of turns increases, a higher inductance can be achieved. However, due to the gain-bandwidth relationship, it also narrows the bandwidth of the ASK-modulated signal, preventing sharp transitions between the 'ON' state and the 'OFF' state. Between the trade-off, the number of turns per layer was determined to be 11 with inner dimensions of 9.3 × 2.6 mm. Both the pitch and the line width were 100 μm, which was the minimum value from the factory.

To estimate an inductance of the designed coil, the modified Wheeler formula was used for calculation [7]. The inductance can be obtained from the formula as follows:

$$L = K_1 \mu_0 n^2 d_{avg} / (1 + K_2 \rho), \tag{2.6}$$

where K_1, K_2 are layout constants, n is the number of turns, μ_0 is the vacuum permeability, $d_{avg} = (d_{out} + d_{in})/2$, and $\rho = (d_{out} - d_{in})/(d_{out} + d_{in})$. Because the formula is based on the assumption of a symmetric coil, such as spiral, square, or hexagonal, the major axis of the designed coil was scaled to a square having the length of the short axis. Then the calculated value was multiplied to the scaled proportion and further multiplied four times for the double layer. Through the calculation, the inductance of the designed coil was obtained as 5.424 μH.

The transmitter coil was also designed and fabricated in a similar way as the receiver coil. The transmitter coil had dimensions of 30 mm × 15 mm with 14 turns at each layer, yielding a calculated inductance of 19.06 μH. The specifications of the designed coil are shown in Table 2.2.

Table 2.2 Dimensions and physical properties of the designed coil antennas

Antenna coil specifications	Tx (intra–oral device)	Rx (implant)
Dimension	30 mm × 15 mm	30 × 6.0 mm
Pitch & Line-width	100 μm	100 μm
Number of turns	28 turns/1 layer	20 turns/2 layer
Simulated inductance	18.43 μH	4.88 μH

2.2.4.5 FEM Simulation

A finite-element method (FEM) simulation was conducted to determine the coupling coefficient between the transmitter antenna and the receiver antenna and the self-inductances of the antennas. Both antennas were modeled into the designed dimensions and two lumped ports were attached to each antenna for a 2-port analysis, as shown in Fig. 2.17. The thicknesses of the metal patterns were not modeled to achieve faster calculations because the thicknesses were small enough compared to the metal width. The models were split into meshes, and the solution was iterated until it converged to the given criteria, namely, the maximum delta S of 0.02. As the calculation result was expressed in terms of port impedances, Z_{11}, Z_{12}, Z_{21} and Z_{22}, self-inductances and the mutual inductance were calculated from the port

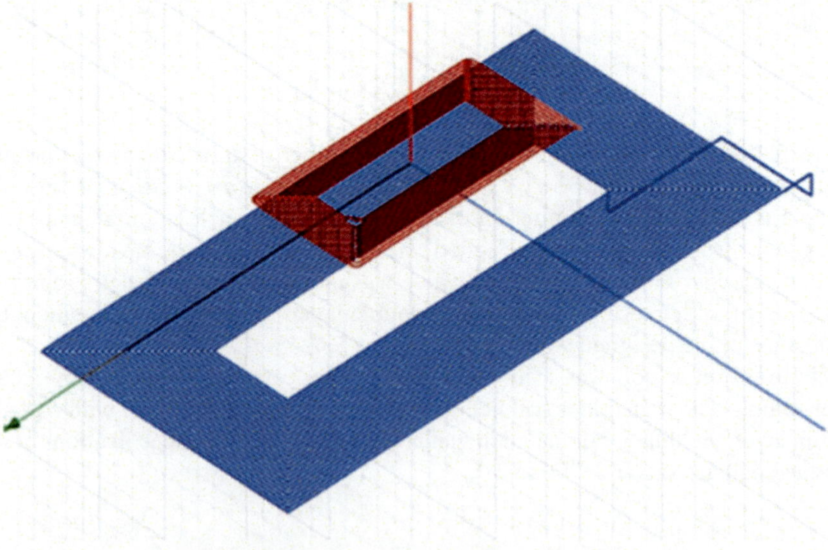

Fig. 2.17 Finite-element method (FEM) simulation drawing using a software HFSS, ANSYS. The receiver and the transmitter coils were moldeled into the designed dimensions and their self inductances and the mutual inductance were calculated from obtained port impedances

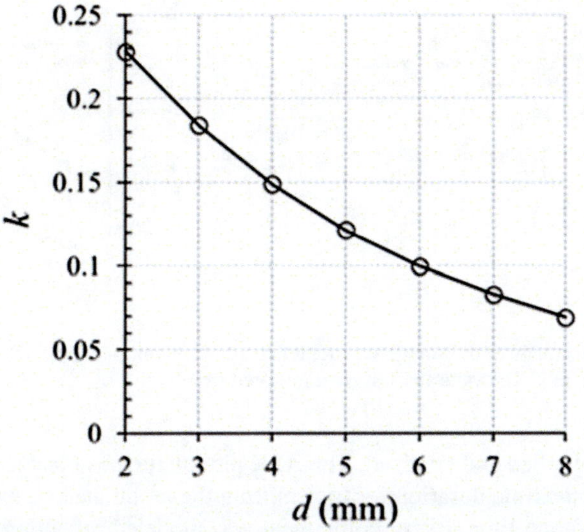

Fig. 2.18 Calculated k plotted over distance from FEM simulation. As expected, k decreases as d increases. The obtained value of k was further used in SPICE simulation

impedances using the following relationships:

$$L_{self} = img(Z_{11\,or\,22})/(2\pi f), \tag{2.7}$$

$$L_{mutual} = img(Z_{12\,or\,21})/(2\pi f), \tag{2.8}$$

where L_{self} and L_{mutual} denote the self-inductance and mutual inductance, respectively, $img(Z)$ denotes the imaginary part of Z, and f is the carrier frequency, which is 2.54 MHz in this application. The coupling coefficient k was calculated from the mutual inductance with the following relationship:

$$k = L_{mutual}/\sqrt{L_T L_R}. \tag{2.9}$$

The obtained self-inductances were compared to the calculated results from (2.7). The coupling coefficient was plotted over the transmission distance ranging from 2 mm to 8 mm as shown in Fig. 2.18.

2.2.4.6 SPICE Simulation

Using the obtained coupling coefficient, the circuit shown in Fig. 2.19a was simulated using a SPICE simulation (OrCad PSPICE 16.6, *Cadence*, USA) to determine

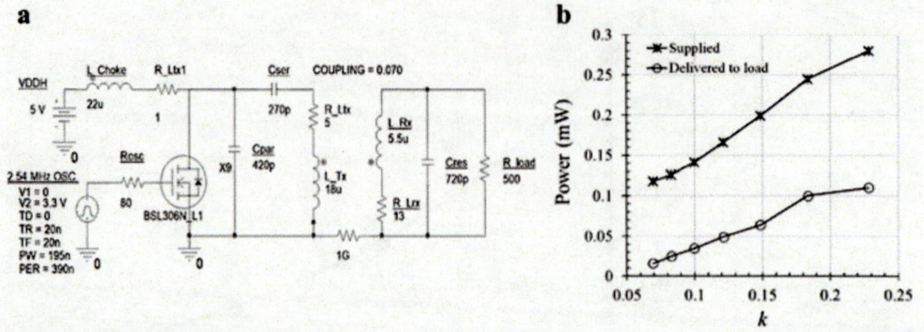

Fig. 2.19 **a** Schematic of the simulated circuit for inductive link **b** results plotting power supplied and delivered over k. The values of k were referenced from Fig. 2.18

the powers supplied and received. The ASK modulator was disabled for the simulation because the time duration for transmitting the modulated signal was negligible over the operation time. Every component was modeled reflecting practical values including vendor-provided IC chip models. The powers at the supply and the load were measured on the telemetry distances. The results are plotted in Fig. 2.19b. Each point of the coupling coefficient k was referenced from the result in Fig. 2.18, which represents a relationship between the telemetry distance and k. For every distance, the supplied power did not exceed 300 mW, which is a marginal limit for around 4-hr operation. As it is expected to operate at the distance of 5 mm in practice, the supplied power of 160 mW gives around 7-hr operation, which is a sufficient operation time for the night hours of sleep. At the time, the power delivered to the load was calculated to be around 50 mW, which was considered sufficient in that most of the supplied power is required for generating biphasic pulses, around 6 mW for our application.

2.2.5 *Evaluations*

2.2.5.1 Wireless Power Transmission

First, the power efficiency of the class-E amplifier was measured. To measure its efficiency, a resistive load was connected after the transmitter antenna to represent the reflecting resistance of the receiver circuit seen from the transmitter. The reflecting resistance can be written assuming a weakly-coupled case ($k \ll 1$) as follows:

$$R_{reflect} = k^2(L_T/L_R)R_L, \tag{2.10}$$

where R_L is the load resistance of the receiver [5]. The values of L_T and L_R at the target frequency were measured using a network analyzer (MS4630B, Amritsu, Japan). The value of R_L not only includes non-linear components, such as diodes and transistors, but also includes active devices of which the power varies over time. To estimate R_L, a test board was fabricated having the same L-C resonator in the receiver circuit and the resistive load attached in parallel to the resonator. The load resistance was found when the voltage across the L-C resonator in the test board was the most similar to that on the actual inductive link and then the value was regarded as the equivalent load resistance of the receiver. From the equivalent resistance, $R_{reflect}$ was calculated. The circuit board having the R-L-C parallel resonance was designed and fabricated to measure the amplifier efficiency (η_{PA}), which can be expressed as

$$\eta_{PA} = \frac{V_{reflect}^2 / 2R_{reflect} + V_{Rtx}^2 / 2R_{Rtx}}{V_S I_S}, \tag{2.11}$$

where $V_{reflect}$ is the voltage across the $R_{reflect}$ and the ground, and V_s and I_s are the voltage and current from the 5 V-power supply. The total power efficiency (η_{tot}) can be obtained using the following equation:

$$\eta_{tot} = \frac{V_{load}^2 / 2R_{load}}{V_S I_S}, \tag{2.12}$$

where V_{load} and R_{load} are the voltage and the resistance across the load, respectively. To further assess the effect of the bio-medium, measurements of the total power efficiency and VDDH (the voltage obtained by rectifying the received signal) were conducted both on the air and the bio-medium.

2.2.5.2 Wireless Data Transmission

In our ASK demodulation strategy based on over-threshold detection, the demodulation result may vary along the distance between the antennas because the magnitude of the received voltage varies. To mitigate the possible errors, the ASIC was programmed to permit PWM signals having duty cycles with one-tenth period of difference. Therefore, it is important to measure duty-cycle errors of the demodulated PWM to determine whether the error is within the error margin over operating distances. To evaluate the errors, around 10^6 bits, composed of '0', '1', and 'F' denoted in 2.2.4.1.1., were transmitted at each distance of 0, 2, 4, 6, and 8 mm. The duty-cycle error of each demodulated bit was obtained through a software code (Matlab R2019a, MathWorks, U.S.) and their averages and standard-deviations were calculated.

2.2.5.3 Electrochemical Measurements in Vitro

The electrochemical impedance spectrums (EISs) and cyclic voltammetries (CVs) were measured on both Au and electrodeposited iridium oxide film (EIROF) using

Solatron 1260 and 1287 systems (Solatron, U.K.) in a three-electrode cell comprised of an Ag/AgCl reference electrode and a mesh-type Pt counter electrode in a PBS solution (Gibco, Thermo Fisher Scientific, U.S). The measurement methods were similar to that in Sect. 2.2.3.1. The EIS was measured for frequencies from 1 Hz to 10 kHz by applying a sinusoidal of 10 mV amplitude without an offset. The CV curve is recorded in the range of −0.6–0.8 V at a scan rate of 1 mV/s at which the CVs of the electrodes can be measured in the steady state. The charge storage capacity (CSC_c) was calculated by integrating the cathodic area of the CV curve using the CorrView software package.

Voltage-transient waveforms of the electrodes were further assessed in a PBS solution to evaluate a charge injection capacity. The three-electrode cell was again constructed and the voltage of the electrode over the reference electrode was measured while applying biphasic current pulses of 1 mA amplitude, pulse rate of 100 Hz, and duration of 540 μs. The current level was increased until the most negative (E_{mc}) and the most positive polarization (E_{ma}) were within the water window, -0.6–0.8 V. The method from the reference [8] was used to figure out E_{mc}, by adding ($\Delta V - V_a$) on a bias level, where ΔV is the voltage transient and V_a is the access voltage.

2.2.5.4 Animal Testing in Vivo

An animal testing was conducted to evaluate the feasibility of the palatal implant system on soft palate stimulation. The preparation procedure of the subject was similar to the method in 2.2.1. A seven-month-old New Zealand white male rabbit was prepared and acclimated to the new environment for the first week. Sedation was induced by intramuscular injection of 0.3 ml/kg of Zoletil (tiletamine hydrochloride, zolazepam hydrochloride; Virbac) and 0.25 ml/kg of Rompun (xylazine; Bayer AG) at the same time. After sleep was induced, an incision was made along the lateral axis of the hard palate, to obtain enough space for the implant to be implanted. Then the implant was inserted toward the soft palate along the incised space. The implantation procedure is shown in Fig. 2.20. After the insertion, the wound was closed by suturing. To deliver power and data to the implant, a transmitter coil was aligned and connected to a board of the intra-oral device. Commands to generate current stimuli having parameters of pulses of 1 mA amplitude, pulse rate of 100 Hz, and duration of 540 μs were transmitted from an external controller via ZigBee. Simultaneously, snapshots of the upper-airway were obtained using C-arm fluoroscopy (Phillips, Netherlands) in both cases where the stimuli were present and absent. All the experimental procedures were approved by the International Animal Care and Use Committee in Seoul National University Bundang Hospital.

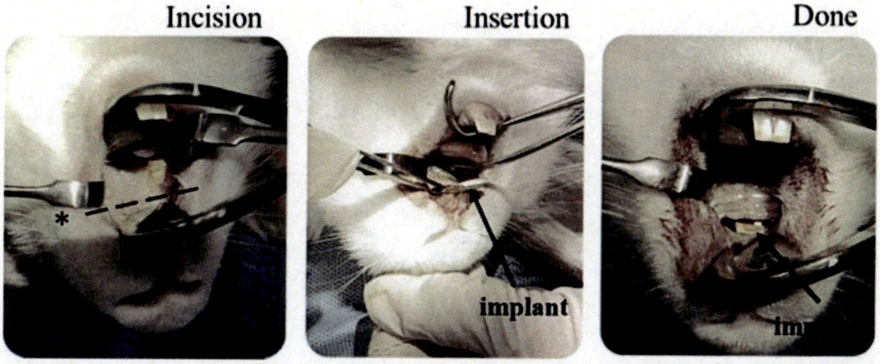

Fig. 2.20 Insertion procedure of the palatal implant. First, a lateral incision was made on the hard palate. Then the implant was inserted from the tips. After enough insertion, implantation of the device was finished

References

1. Bard AJ, Faulkner LR (2001) Electrochemcal methods: fundamentals and Appl, 2nd edn. Wiley, New York
2. Min KS, Oh SH, Park MH, Jeong J, Kim SJ (2014) A polymer-based multichannel cochlear electrode array. Otol Neurotol 35(7):1179–1186. https://doi.org/10.1097/MAO.0000000000000292
3. Shin S, Kim J, Jeong J, Gwon TM, Choi GJ, Lee SE, Kim J, Jun SB, Chang JW, Kim SJ (2016) High charge storage capacity electrodeposited iridium oxide film on liquid crystal polymer-based neural electrodes. J Sens Mater 28(3):243–260
4. Sokal NOJQ (2001) Class-E RF power amplifiers. J QEX 204(1):9–20
5. Harrison RR (2007) Designing efficient inductive power links for implantable devices. In: 2007 IEEE international symposium on circuits and systems. IEEE, pp 2080–2083
6. Vaillancourt P, Djemouai A, Harvey J, Sawan M EM radiation behavior upon biological tissues in a radio-frequency power transfer link for a cortical visual implant. In: Proceedings of the 19th Annual International Conference of the IEEE Engineering in Medicine and Biology Society.'Magnificent Milestones and Emerging Opportunities in Medical Engineering'(Cat. No. 97CH36136), 1997. IEEE, pp 2499–2502
7. Mohan SS, del Mar Hershenson M, Boyd SP, Lee TH (1999) Simple accurate expressions for planar spiral inductances. J IEEE J Solid-State Circuits 34(10):1419–1424
8. Cogan SF (2008) Neural stimulation and recording electrodes. Annu Rev Biomed Eng 10:275–309. https://doi.org/10.1146/annurev.bioeng.10.061807.160518

Chapter 3
Results

In this chapter, experimental results of the evaluations described in Methods are provided with figures and graphs with analyses. Fabrication results, in vitro and in vitro experimental results with brief analyses are provided for each system.

3.1 Magnetic Cuff System for HNS

3.1.1 Fabricated System

Pictures of the fabricated MENCE is in Fig. 3.1. The MENCE has a pair of stimulation electrodes for a neural interface and connector pads at the other end for access to an external pulse generator (Fig. 3.1a). The stimulation site area is 5.6 mm^2 with a width of 3.5 mm and a height of 1.6 mm. The internal cuff radius is 1.27 mm, which targets a hypoglossal nerve (Fig. 3.1b). The measured magnetic flux density of the magnet is 0.112 T, which is lowered by 20% compared to that before the lamination process, but still strong enough for closing. The external pulse generator is shown in Fig. 3.2. A 9 V battery provides power to the PCB and a user can set parameters using the embedded switches. The generated current stimuli are delivered to the electrodes through a FFC connector.

3.1.2 Electrochemical Measurements in Vitro

Figure 3.3a shows the EIS result performed on the fabricated electrodes (n = 10). The measured impedance is $0.638\angle{-67.8^\circ}$ kΩ ($\pm 0.210\angle{-3.27^\circ}$ kΩ) at 1 kHz and $5.27\angle{-82.1^\circ}$ kΩ ($\pm 0.694\angle{-1.07^\circ}$ kΩ) at 100 Hz. The average values of access

J. Seo, *Development of Implantable Electronics as Novel Approaches to Obstructive Sleep Apnea*, Springer Theses, https://doi.org/10.1007/978-981-15-8327-8_3

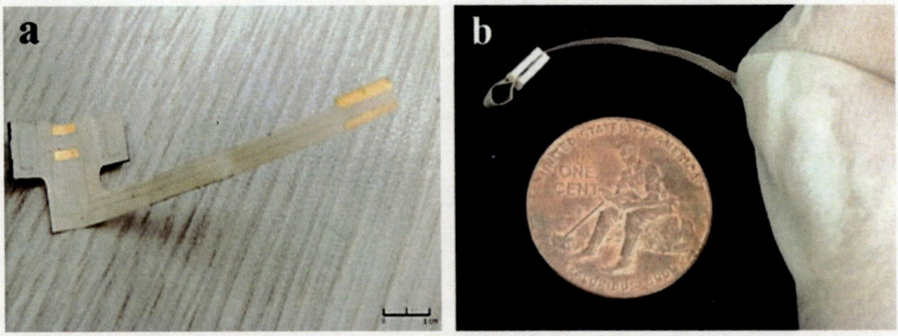

Fig. 3.1 **a** Fabricated magnet-embedded nerve cuff electrode (MENCE) and **b** its side view in closed form

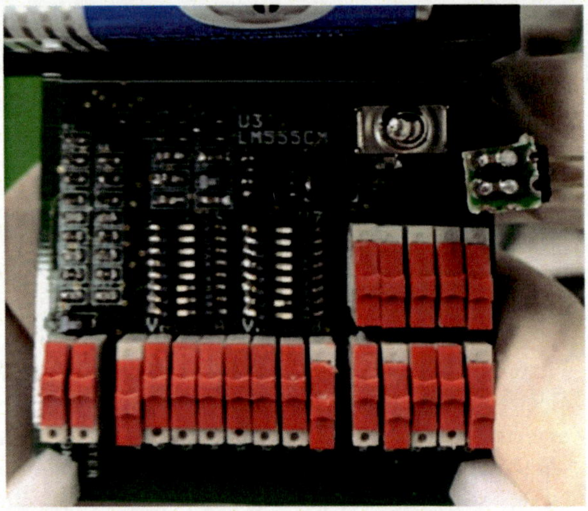

Fig. 3.2 The external pulse generator fabricated on PCB. Mechanical switches were soldered on the board to set stimulation parameters

resistance and cut-off frequency are 0.145 kΩ (±1.74 Ω) and 3.98 kHz (±3.84 Hz), respectively. The measured magnitude of the impedance is lower than those of prior arts. The reason can be found from the electrode design, the impedance of the electrode decreases as the site area increases [1, 2]. A representative CV curve of the electrode is plotted in Fig. 3.3b. The average value of the CSCc is 1.69 mC/cm^2 (±0.602 mC/cm^2). The measured $\mathrm{CSC_c}$ implies that the electrode can deliver a charge up to 94.6 μC for the site area of 5.6 mm^2, which is enough value for the required amount of charge of 13 nC for our application.

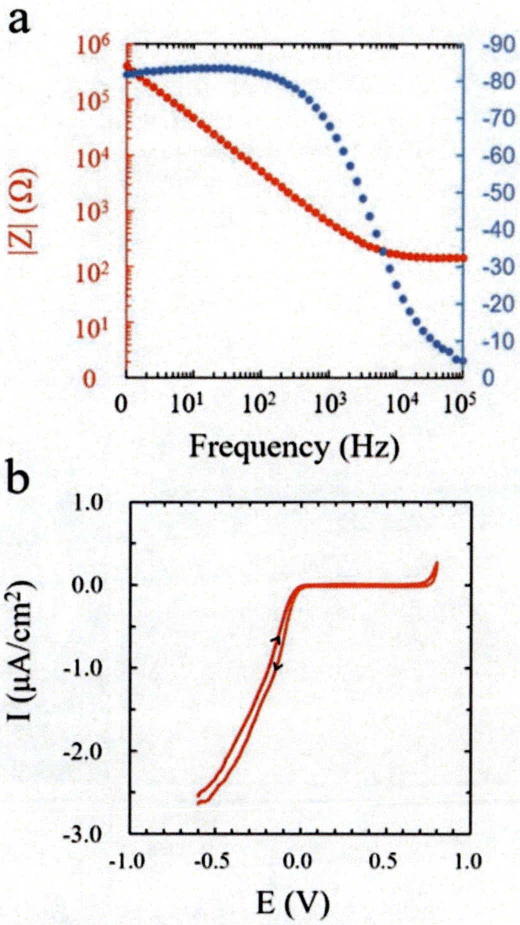

Fig. 3.3 Results of in vitro assessments of MENCE **a** absolute value and phase degree of the impedance, **b** cyclic voltammogram

3.1.3 *Animal Testing in Vivo*

In vivo results of the hypoglossal nerve stimulation are shown in Fig. 3.4. Tongue protrusion was clearly observed with bare eyes during the stimulation (Fig. 3.4a). The minimum level for the stimulation was measured to be 41.0 μA, inducing a mild reaction. When the level was over over 53.2 μA, strong protrusion of the tongue was observed. Considering that the stimulation should not occur arousal during sleep, the mild stimulation will be desirable for a practical application. Further research will be necessary for finding an optimal stimulation level. Figure 3.4b shows C-arm fluoroscopy snapshots of the experiment. In the figures, the changed position of the cuff electrode is seen due to the strong contraction of the genioglossus. Figure 3.4c shows EMG signal recorded at the neck at the same time. A pair of peaks were periodically measured in EMG. Each large peak represents the start of the stimulation

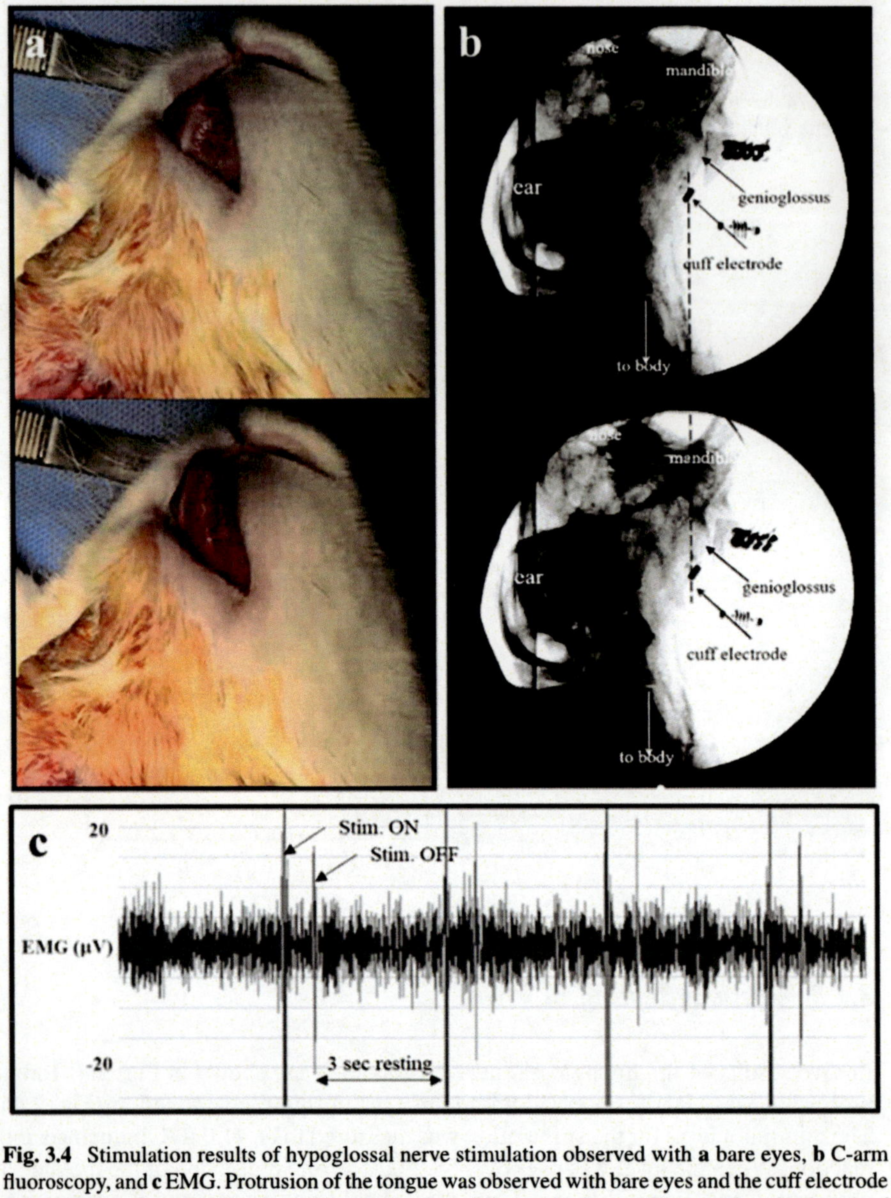

Fig. 3.4 Stimulation results of hypoglossal nerve stimulation observed with **a** bare eyes, **b** C-arm fluoroscopy, and **c** EMG. Protrusion of the tongue was observed with bare eyes and the cuff electrode was forced to the mandible due to the contraction of the genioglossus. The synchronized peak of EMG supports the evidence of the contraction

and small peak represent the end of the stimulation, which are due to extraction and release of the stimulation, respectively. The synchronized EMG peaks shows the evidence of the genioglossus contraction.

3.2 Palatal Implant System for SPS

3.2.1 Pilot Study in Vivo

Current stimuli with varied pulse frequency were delivered to the soft palate of the subject. Strong contraction of the soft palate was observed within the frequency range of 70–120 Hz. Mild responses were in the range of 130–150 Hz but no responses under 60 Hz and over 160 Hz. Based on observation of the contraction, the threshold level was 1.10 mA with duration and frequency of 636.3 μs and 100 Hz, respectively. Figure 3.4 shows the upper airway snapshot recorded using a C-arm fluoroscopy. Both snapshots were taken at points when the soft palate was maximally relaxed. Compared to the untreated case, the elevation of the soft palate was detected when the stimulation was applied, as shown in Fig. 3.5. From the result, it can be deduced that the current stimulation on the soft palate has an effect on broadening the retropalatal region.

To demonstrate a long-term reliability of the implant, a mock-up was fabricated as shown in Fig. 3.6a. The mock-up has a dimension of 21 mm × 10 mm with a thickness of 2 mm. The two shanks, each with 13 mm-long, have 1.5 mm widths with 0.7 mm-thickness. Figure 3.6b–d shows the insertion process of the mock-up. There has been no evidence of any adverse events since the mock-up was inserted.

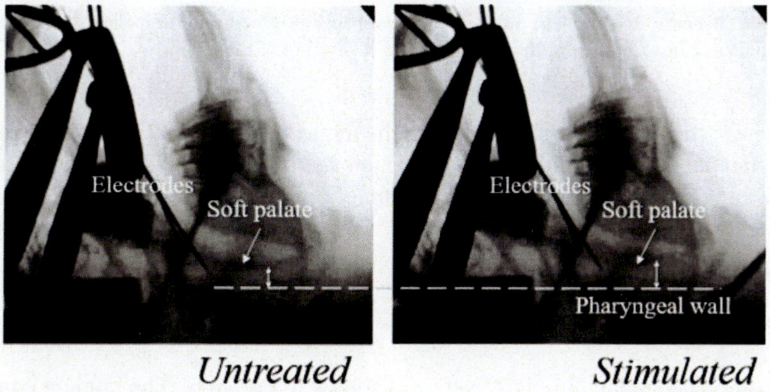

Fig. 3.5 Snapshots of the upper airway without stimulation (top) and with stimulation (bottom) recorded via C-arm fluoroscopy. The soft palate was elevated when the stimulation was applied, resulting in retropalatal opening

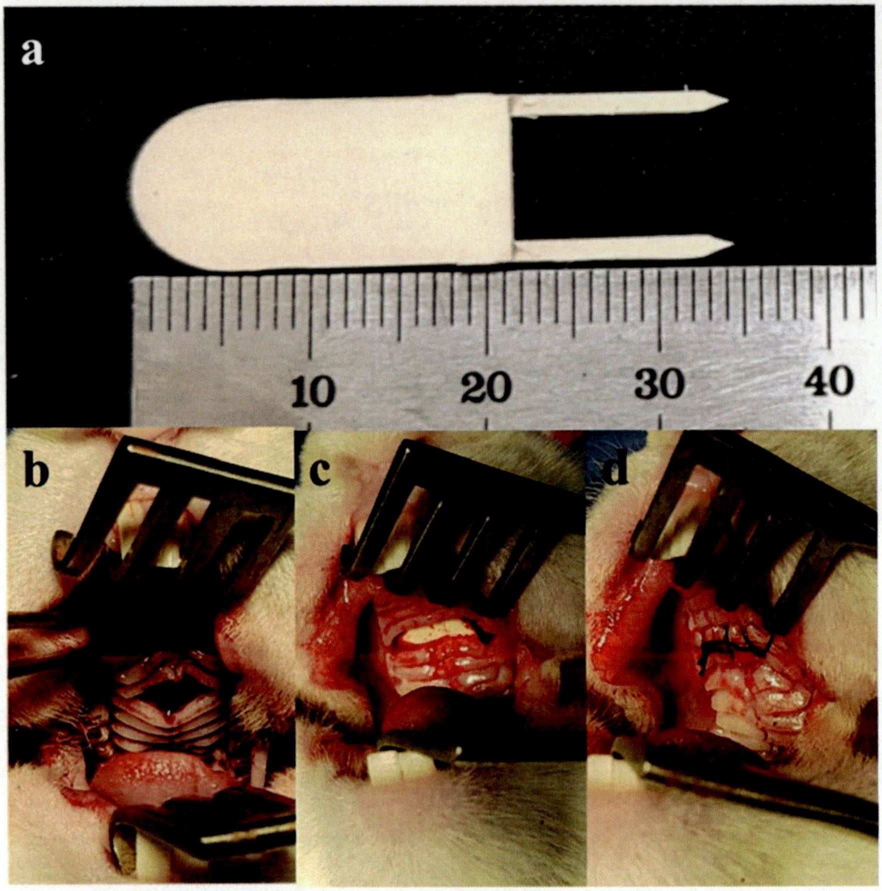

Fig. 3.6 **a** Fabricated implant mock-up and its insertion process to the palate. **b** Incision of the hard palate, **c** implantation of the mock-up, and **d** suturing of the wound

However, the mock-up protruded from the incision point after the two months of the implantation. We thought that the protrusion was due to the bulky size of the implant, therefore decided to reduce the dimension of the implant.

3.2.2 Fabricated System

The fabricated palatal implant system is shown in Fig. 3.7a. The implant consists of multiple LCP layers including the functional layers of the coil, electrodes facing top and bottom, and adhesion layers (Fig. 3.7b). The fabricated implant had dimensions of 19 mm × 8.0 mm with a thickness of 3 mm including the encapsulation silicone

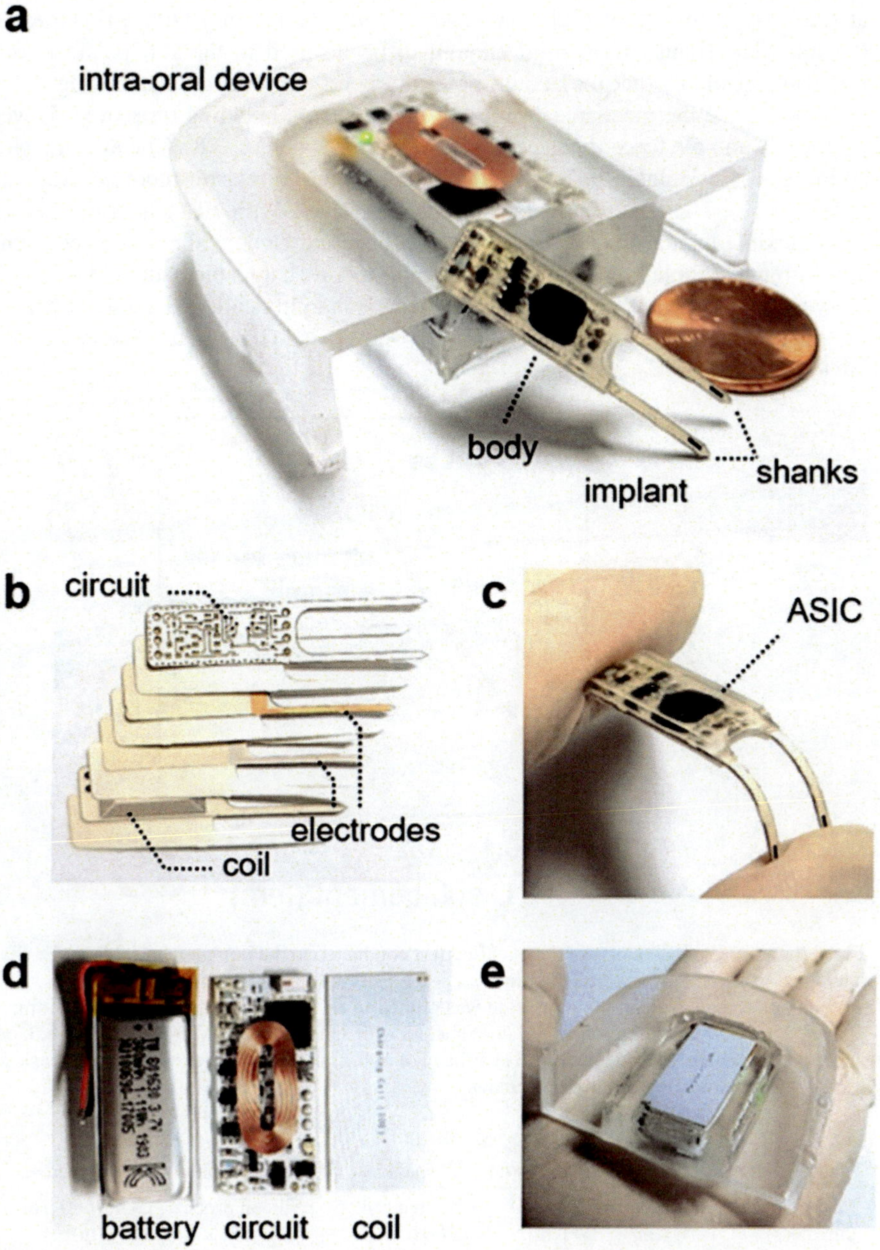

Fig. 3.7 Fabricated palatal implant system. **a** Intra-oral device and palatal implant. **b** Layer configurations of the implant (unnoted layers to be adhesion layers). **c** Flexible shanks of the implant. **d** Electronic parts of the intra-oral device, consisting of rechargeable battery, circuit board, and coil. **e** The intra-oral device seen at the other side

at the body. The size of each shank was 13 mm × 1.5 mm with a thickness of 0.7 mm. The shanks were rigid enough to be inserted to the soft palate muscle but also flexible so that they could be easily bendable with the hands (Fig. 3.7c). To characterize the mechanical property of the shank, buckling tests (n = 5) were performed and the force-displacement plot was drawn (Fig. 3.8). The average yield strength was calculated to be 2.51 N (±0.32), and the elastic modulus was 1.32 GPa (±0.17). The result indicates that the LCP shank with 700-μm thickness was more flexible than a tungsten rod, silicon probe, and COP, while it showed higher yield strength (Table 3.1) [3, 4]. A lithium-polymer rechargeable battery (Tw801630, Taiwoo, China; 1.11 Wh), a circuit board with a charging coil, and a transmitter coil are the electronic components of the intra-oral device (Fig. 3.7d). The device was

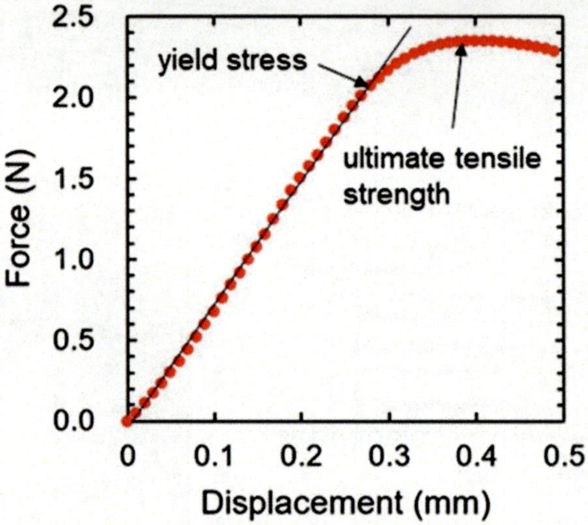

Fig. 3.8 A representative plot of the buckling test conducted to the implant. One side of the shank was fixed to a force sensor attached to a data acquisition (DAQ) unit (Nano 17, ATI Industrial Automation Inc., U.S.) and the other side was pinned to an acrylic plate placed on the bottom. A single-axis oil hydraulic micromanipulator attached with a motor driver (MO-81, Narishige, Japan) and the force sensor, moved downward at a speed of 100 μm/s. The force exerted on the shank was recorded by the function of the displacement

Table 3.1 Result of the buckling tests on LCP shanks and its comparison to previous studies

Material (width x thickness)	LCP shank (1200 μm × 700 μm)	Tungsten rod (ϕ = 200 μm)	Silicon probe (100 μm × 100μm)	COP (650 μm × 200 μm)
Elastic modulus (GPa)	1.322 (n = 5)	5.868 (n = 5)	2.964 (n = 5)	2.001 (n = 4)
Yield strength (< N)	2.51 ± 0.32	0.23 ± 0.11	0.27 ± 0.1	0.22 ± 0.008

Table 3.2 Properties of fabricated antenna coil. Absolute value (red) and phase degree (blue) of the electrode impedance (left), cyclic voltammogram of the electrode (right)

	Dimension	Turns	L (μH)	R (Ω)	Q_L
Transmitter	30 mm × 15 mm	28(2 layers)	17.50	16.30	16.86
Receiver	13 mm × 6.0 mm	20(2 layers)	5.475	13.05	2.990

able to operate for 6 h with heat dissipation of 2 °C increase in the soak test, which will be discussed in 4.2.1. Figure 3.7e provides the opposite view of the IOD from Fig. 3.7a.

The properties of the fabricated coil antennas are presented in Table 3.2. The dimensions of each coil were determined based on the constraint of the biological dimensions. The quality factor, Q, of each coil was intentionally lowered to obtain enough bandwidth for ASK modulation. As the gain and the bandwidth are inversely proportional to each other due to the gain-bandwidth product, smaller Q was selected at the expense of lowered power efficiency.

3.2.3 *Wireless Power Transmission*

First, coupling coefficient k was measured by varying the distance between the coils. The value of k was measured to be in the range of 0.050–0.208, which is roughly 22% lower than the simulation results in 2.2.5.2 at distances of 2–8 mm. The discrepancy between the measurement and simulation results might be due to misalignment in the measurement and simplified model of the simulation. The received peak-to-peak voltage at the coil, V_{Rpp}, has the following proportional relationship with k:

$$V_{Rpp} = Ak\sqrt{\frac{L_R}{L_T}}Q'_T Q'_R V_s, \tag{3.1}$$

where A is the voltage gain of the amplifier; L_R and L_T are inductances of receiver and transmitter, respectively; Q'_T and Q'_R are the loaded $Q's$ of the receiver and the transmitter, respectively; and V_s is the supply voltage. The coupling coefficient k and the received peak-to-peak voltage V_{pp}, are plotted in Fig. 3.9a showing a linear relationship. The minimum value of V_{pp} was 5.9 V, which is still a valid compliance considering that the required minimum compliance voltage was 2 V in a PBS solution for our 1-mA stimulation, and the compliance in vivo was reported to be about 3 times higher than in vitro [5].

The reflected resistance was calculated from Eq. (2.10) using the obtained k, and the amplifier efficiency was measured at each distance. The averaged amplifier efficiency was 65.6% (Fig. 3.9b), which is lower than routinely reported efficiencies of 85% from other studies [6]. This is due to relatively low Q'_T because in our application both power and data transmission occur.

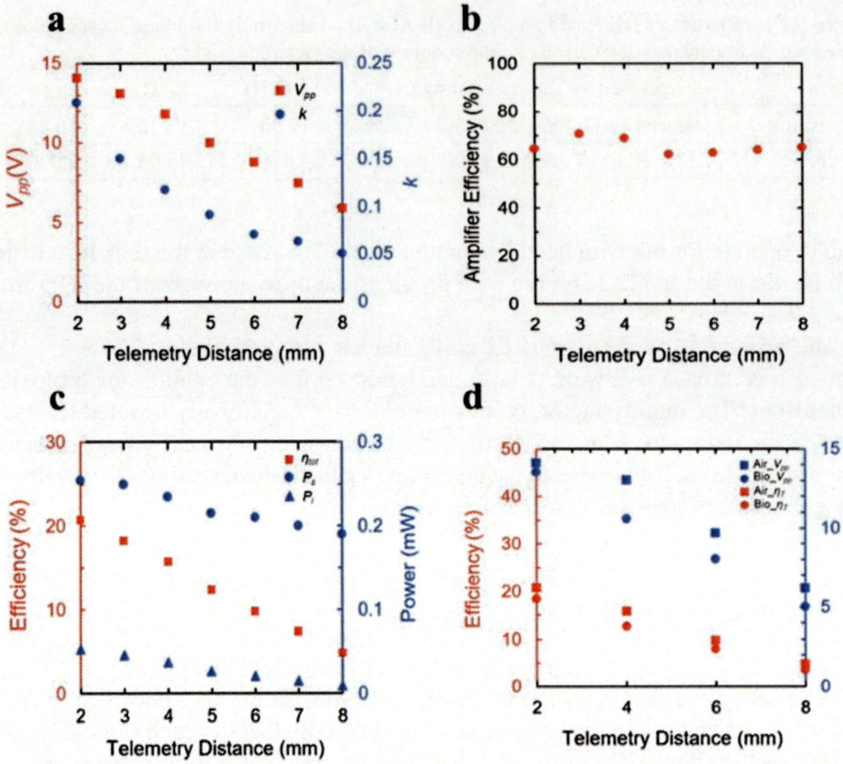

Fig. 3.9 Power transmission results of the palatal implant system varying telemetry distance ranging 2–8 mm. **a** VDDH and k **b** measured amplifier efficiency **c** total power efficiency and power of supply and load **d** comparison of air and bio-medium in terms of VDDH and total power efficiency

Power supplied by the source in the intra-oral device and power delivered to the load in the implant are represented by blue dots in Fig. 3.9c. The maximum supplied power was 255 mW, which allowed 4.35 h of operation. The measured total power efficiency, P_s/P_l, is represented by red dots in Fig. 3.9c. The efficiency was 4.87–20.7% in the telemetry distance. At the distance of 8 mm, which was the maximum distance over which the circuit in the implant operated, the load power was measured to be 10 mW. Figure 3.9d shows a comparison plot of the air and the bio-medium in terms of V_{pp} and the total efficiency η_T with distance variation. Both V_{pp} and η_T were lower in the bio-medium than the air as expected; the decreases were not significant but had a minor effect of decreased compliance.

3.2.4 Wireless Data Transmission

The result of the duty-cycle measurement is shown in Fig. 3.10. Every bit, 0, 1, and F, represented as the duty-cycle of PWM, was measured to have an averaged error

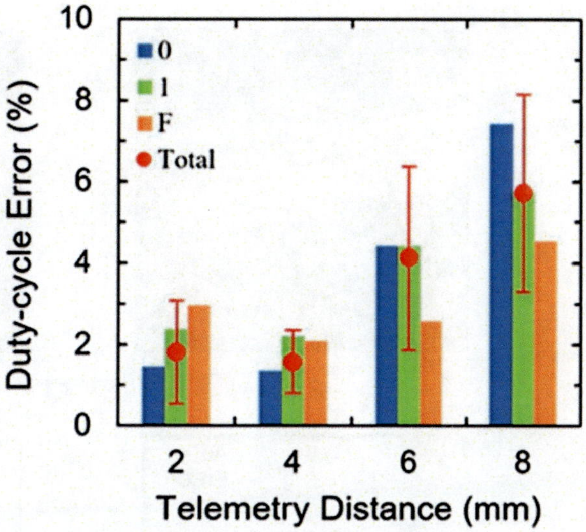

Fig. 3.10 Measured telemetry error in terms of duty-cycle. Error values of each bit and their averages are plotted over the telemetry distance

within the margin of 10%, at all telemetry distances ranging from 2 to 8 mm. The error increased at telemetry distances over 4 mm, where the large drop of V_{pp} in air was observed, as shown in Fig. 3.9d. It seems that the decreased signal level affected the performance of the ASK demodulator. However, the errors were in tolerance, implying successful data transmission at the tested distances.

3.2.5 Electrochemical Measurements in Vitro

The averaged EIS results (n = 8) of the fabricated Au and EIROF electrodes are shown in Fig. 3.11, representing the averaged magnitude and phase of the measured impedances with their standard deviation noted. The low cut-off of EIROF is due to EIROF's high porosity which is a highly desirable property for a stimulation electrode [7]. The magnitude of EIROF was measured to be 390 Ω at 1 kHz, which is nine times lower than that of Au, 4.13 kΩ. The EIROF already exhibited a near-resistive phase, while Au stayed capacitive at 1 kHz. The high porosity of the EIROF electrode contributed to a higher charge injection/storage capacity, as shown in Fig. 3.12.

Figure 3.13 shows a representative plot of cyclic voltammetry conducted on Au and EIROF. The averaged cathodal charge storage capacity (CSC_c) of EIROF was 62.25 mC/cm^2 with the standard deviation of 19.17 mC/cm^2, while it was 0.7019 mC/cm^2 with 0.2695 mC/cm^2 for Au. EIROF having a porous surface had a higher electrochemical surface area (ESA) per unit geometric surface area (GSA); therefore,

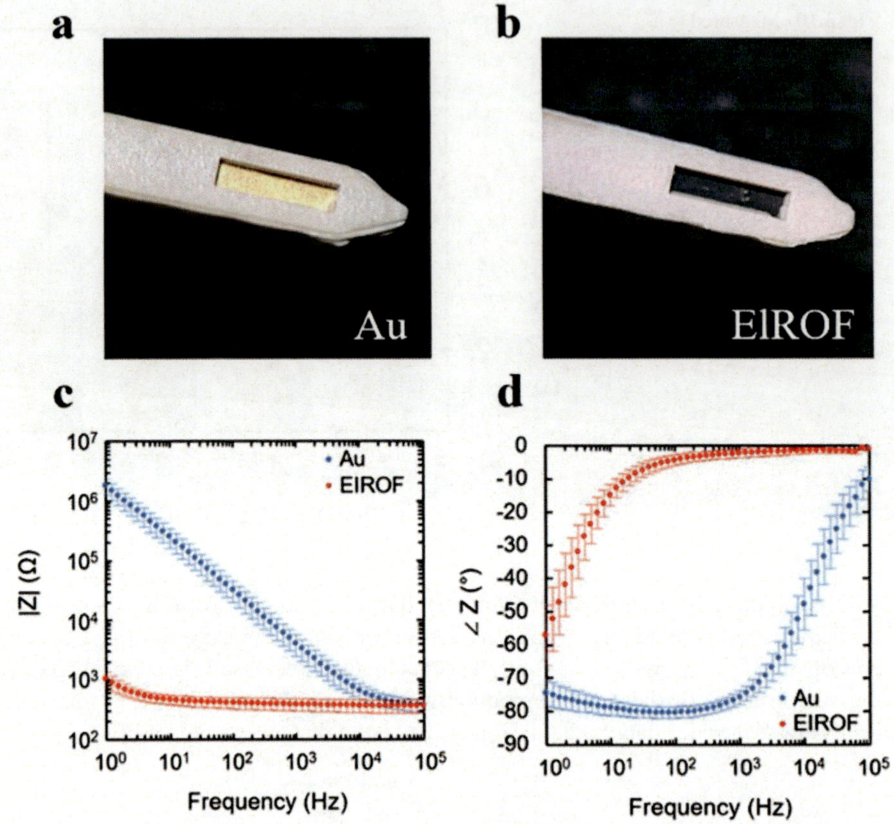

Fig. 3.11 Fabricated **a** Au and **b** EIORF electrode image and results of EIS. The impedance of EIROF was nine-times lower than that of Au. **c** The magnitude in logarithmic scale. **d** Phase in degree. EIROF exhibits resistive impedance at 1 kHz while Au remains capacitive

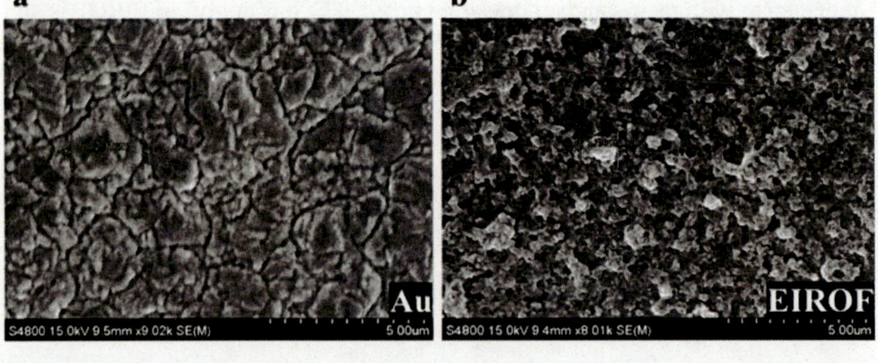

Fig. 3.12 Scanning electron microscopy (SEM) images of **a** Au and **b** EIROF surface. Both images are taken at the same magnification (S-4800, Hitachi, Japan)

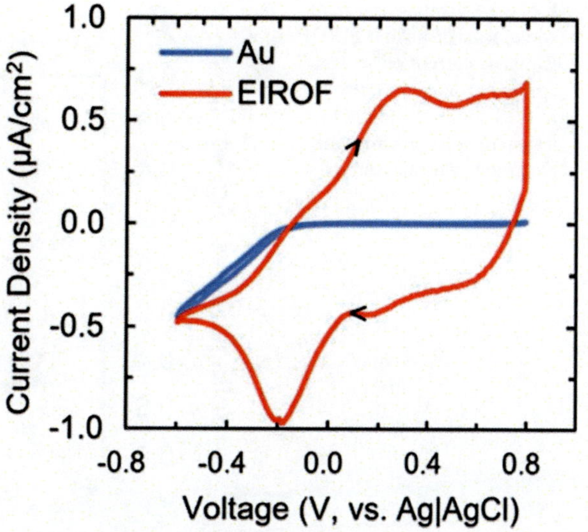

Fig. 3.13 Representative plots of the cyclic voltammetry of Au (blue) and EIROF (red). The plots were extracted from fifth cycle of the voltammetry. CSC_c of EIROF was measured to be 88.9-fold higher than that of Au

it exhibits a CSC_c about 90 times higher than that of Au. The amount of cathodal charge used in this experiment was 270 nC (1 mA × 270 μs). Therefore, both CSC_c's of Au and EIROF were measured to be enough for the stimulation.

The voltage transients of Au and EIROF are plotted in Fig. 3.14a and b, respectively. From the bias level (V_{ipp}), an electrode faced iR drop immediately after current pulse was applied. The amount of the potential drop is called an access voltage, which includes the access resistance of the electrode and electrolyte. This voltage drop should be excluded for evaluating a charge injection capacity as they do not participate in the polarization across the electrode-electrolyte interface. Thus, the most negative polarization E_{mc} for cathodal phase can be calculated as

$$E_{mc} = E_{ipp} + (V_{drv} - V_a), \tag{3.2}$$

where V_{drv} is the driving voltage in Fig. 3.12 [7, 8]. The most positive polarization (E_{ma}) can be similarly obtained using Eq. (3.2) for the anodic phase. Both E_{ma} and E_{mc} must not exceed a voltage range at which irreversible faradaic reaction occurs, namely, –0.6 to 0.8 V (vs. Ag|AgCl reference) for the water limit. As seen in Fig. 3.14, both Au and EIROF do not exceed the water limit under the stimulation parameters of 1-mA amplitude, 540-μs duration, and 100-Hz pulse rate. The polarization voltage across the interface, V_p, was drastically decreased for EIROF due to its low impedance, complying with the result in Fig. 3.11. Figure 3.15a shows the transient result with current variation from 0.4 to 2.0 mA. The values of E_{mc} were far less than the cathodic water limit of –0.6 V, as can be seen from the plot.

To determine the charge injection limits of Au and EIROF, the voltage transients of Au and EIROF were measured and plotted in $\underline{V_{drv}}$ and E_{mc} as a function of charge density (Fig. 3.15b). The driving voltages of both Au and EIROF increased as the

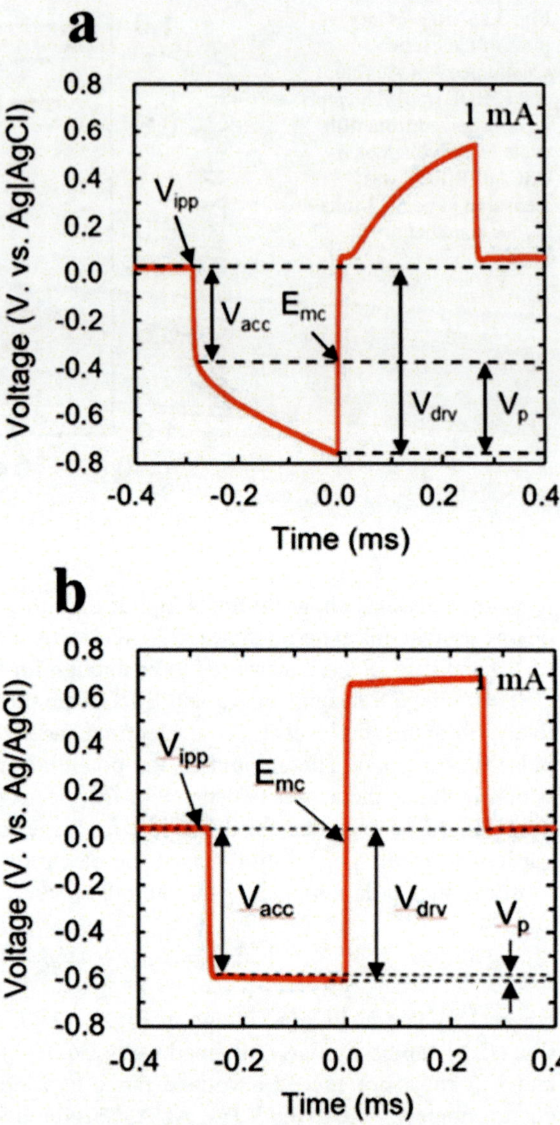

Fig. 3.14 Voltage transient waveforms of **a** Au **b** EIROF. Biphasic current pulse was applied to soaked electrodes in PBS solution and the transient voltage across the electrodes was measured

charge density increased, but E_{mc} of EIROF remained relatively constant compared to E_{mc} of Au. This implies that V_a takes up most of the driving voltage for EIROF, and the polarization across the interface is very low. Au exceeded the cathodic water limit when a charge density of 0.054 mC/cm^2 was applied, while EIROF still remained below the limit. EIROF exceeded the water limit at 0.162 mC/cm^2, when the current level of 6 mA was applied. From the voltage transient results, it was shown that EIROF has a charge injection limit approximately three times higher, and both Au and EIROF were safe enough for the charge density of 0.027 mC/cm^2.

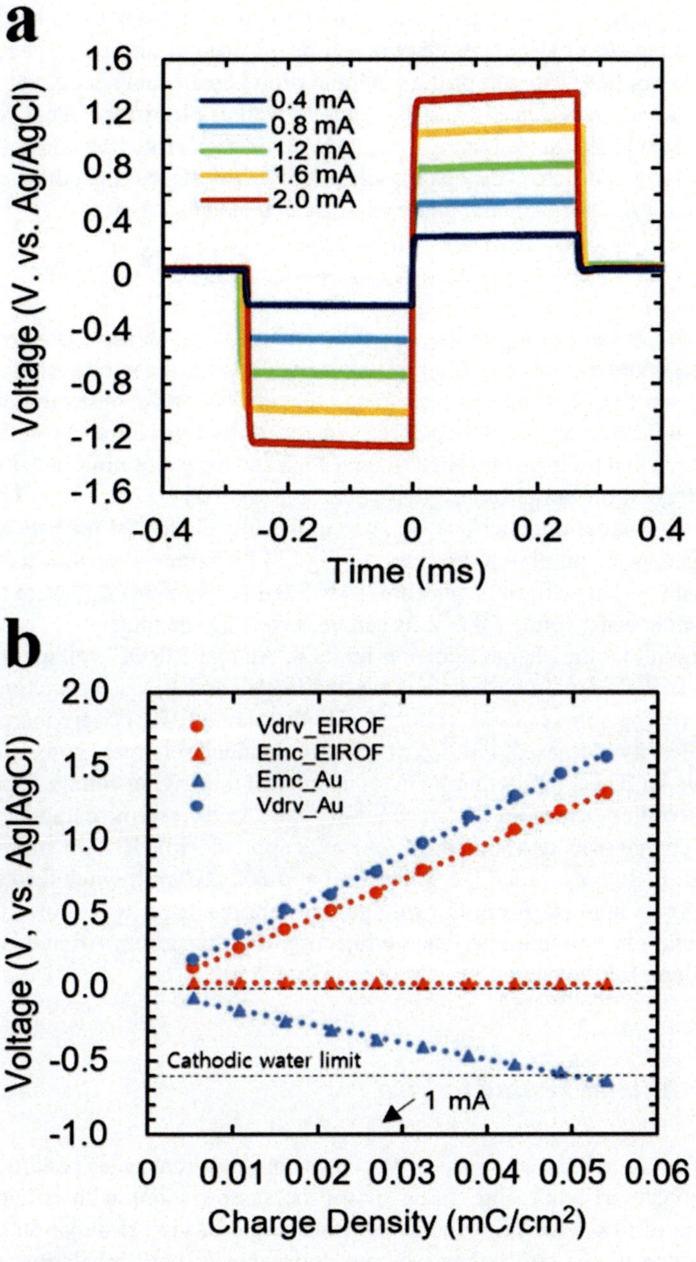

Fig. 3.15 **a** Voltage transient waveforms of EIROF by varying current level from 0.4 to 2 mA. E_{mc} was not exceeded even at 2.0 mA. **b** Plots of V_{drv} and E_{mc} of EIROF and Au versus charge density. Au exceeded the water limit at 0.054 mC/cm^2

Voltage transients of Au and EIROF are plotted in Fig. 3.14a and b, respectively. From the bias level (V_{ipp}), an electrode faces *iR* drop at immediately after current pulse was applied. Amount of the potential drop is called an access voltage, which contains access resistance of the electrode as well as electrolyte. This voltage drop should be excluded for evaluating a charge injection capacity as they do not participate in the polarization across the electrode-electrolyte interface. Thus, the most negative polarization E_{mc} for cathodal phase can be calculated as follows

$$E_{mc} = E_{ipp} + (V_{drv} - V_a) \tag{3.2}$$

where V_{drv} is the driving voltage in Fig. 3.12 [7, 8]. The most positive polarization (E_{ma}) can be similarly obtained using Eq. (3.2) for the anodic phase. Both E_{ma} and E_{mc} must not exceed a voltage range that occurs irreversible faradaic reaction, namely, −0.6–0.8 V (vs. Ag|AgCl reference) for the water limit. From Fig. 3.14, it can be seen that both Au and EIROF do not exceed the water limit under stimulation parameter of 1 mA amplitude, 540 μs duration, and 100 Hz pulse rate. The polarization voltage across the interface, V_p, was drastically decreased for EIROF due to its low impedance, complying the result in Fig. 3.11. Figure 3.15a shows the transient result with varying current level from 0.4 to 2.0 mA. Values of E_{mc} were far less than the cathodic water limit, −0.6 V, as can be seen from the plot.

To figure out the charge injection limits of Au and EIROF, voltage transients of Au and EIROF were measured and plotted in $\underline{V_{drv}}$ and E_{mc} as a function of charge density (Fig. 3.15b). Driving voltages of both Au and EIROF were increased as the charge density increased, but E_{mc} of EIROF remained relatively constant compared to E_{mc} of Au. This implies that V_a takes up most of the driving voltage for EIROF and the polarization across the interface is very low. Au exceed the cathodic water limit when a charge density of 0.054 mC/cm^2 was applied while EIROF still remained in the limit. EIROF exceeded the water limit at 0.162 mC/cm^2, when the current level of 6 mA was applied. From the voltage transient results, it was shown that EIROF has around three-time higher charge injection limit, and both Au and EIROF were safe enough for the charge density of 0.027 mC/cm^2.

3.2.6 *Animal Testing in Vivo*

Figure 3.16a and b shows upper-airway snapshots under untreated conditions (inhalation, exhalation) while Fig. 3.16c shows the same regions with stimulation. The untreated results were provided as controls of the in vivo stimulation assessment. Any change of the soft palate was not noticeable in both inhalation and exhalation under the untreated conditions. On the other hand, distinctive deformation of the soft palate was observed when stimulation was applied, leading to a broadened retropalatal region. Further measurements to demonstrate the therapeutic effect of stimulation could not be performed because the subject was not OSA-modeled in the

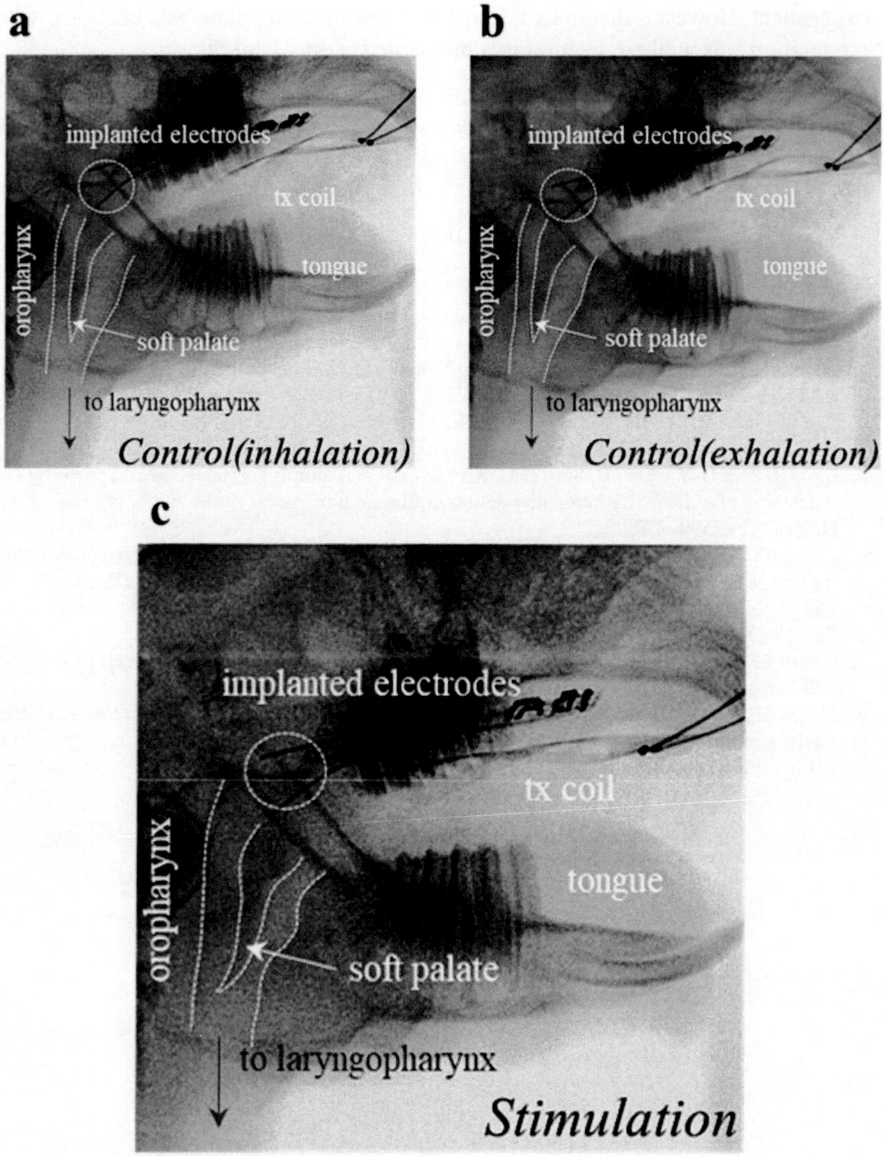

Fig. 3.16 Snapshots of the upper-airway observed from C-arm fluoroscopy. **a** Inhalation with no stimulation. **b** Exhalation with no stimulation **c** Stimulation applied. Distinctive constraint of the soft palate was observed during stimulation compared to the controls, **a**, **b**

experiment. However, distinctive deformation of the soft palate was observed when stimulation was applied, leading to a broadened retropalatal region.

References

1. Shin S, Kim J, Jeong J, Gwon TM, Choi GJ, Lee SE, Kim J, Jun SB, Chang JW, Kim SJ (2016) High charge storage capacity electrodeposited iridium oxide film on liquid crystal polymer-based neural electrodes. J Sensor Mater 28(3):243–260
2. Seo J, Choi GJ, Park S, Lee J, Baek C, Jang J, Lim J, Shin S, Seo K, Seo JM (2017) Wireless navigation of pigeons using polymer-based fully implantable stimulator: a pilot study using depth electrodes. In: 2017 39th annual international conference of the IEEE engineering in medicine and biology society (EMBC). IEEE, pp 917–920
3. Lee SE, Jun SB, Lee HJ, Kim J, Lee SW, Im C, Shin H-C, Chang JW, Kim SJ (2012) A flexible depth probe using liquid crystal polymer. J IEEE Trans Biomed Eng 59(7):2085–2094
4. Shim S, Park H-Y, Choi GJ, Shin H-C, Kim SJ (2019) A simply fabricated neural probe by laser machining of a thermally laminated gold thin film on transparent cyclic olefin polymer. J ACS Omega 4(2):2590–2595
5. Leung RT, Shivdasani MN, Nayagam DA, Shepherd RK (2014) In vivo and in vitro comparison of the charge injection capacity of platinum macroelectrodes. J IEEE Trans Biomed Eng 62(3):849–857
6. Sokal NOJQ (2001) Class-E RF power amplifiers. J QEX 204(1):9–20
7. Cogan SF (2008) Neural stimulation and recording electrodes. Annu Rev Biomed Eng 10:275–309. https://doi.org/10.1146/annurev.bioeng.10.061807.160518
8. Cogan SF, Plante T, Ehrlich J (2004) Sputtered iridium oxide films (SIROFs) for low-impedance neural stimulation and recording electrodes. In: The 26th annual international conference of the IEEE engineering in medicine and biology society. IEEE, pp 4153–4156

Chapter 4
Discussions

Based on the results, further analyses are presented and discussed in relation to safety issues, the reliability of the systems, potential developments, and so forth. They include theoretical calculation, a simple reliability test in vitro, analysis of the results based on anatomy, and suggestions for future work.

4.1 Magnetic Cuff System for HNS

4.1.1 Cuff-to-Nerve Diameter Ratio

A mismatch between curled radius of the MENCE and the nerve may induce critical problems such as necrosis due to reduced blood flow. Therefore, it is necessary to evaluate whether the radius of the MENCE does not interfere blood flow of the nerve. According to Naples et al., the cuff-to-nerve diameter ratio (CNR) can play such a role, and the relationship between a change in the internal radius and the pressure can be expressed as follows [1].

$$\Delta \mathrm{r} = \frac{r_c \Delta P}{E\left[(r_c + h)^2 - r_c^2\right]} * \left[(1 - v) + (1 + v) * \frac{(r_c + h)^2}{r_c^2}\right] \tag{4.1}$$

Δr = Change in the Internal Radius
ΔP = Change in the Internal Pressure
r_c = Internal Cuff Radius
h = Cuff Wall Thickness
E = Young's Modulus of the Substrate
v = Poisson's Ratio of the Substrate

J. Seo, *Development of Implantable Electronics as Novel Approaches to Obstructive Sleep Apnea*, Springer Theses, https://doi.org/10.1007/978-981-15-8327-8_4

Note that the application over 20 mmHg of pressure to a peripheral nerve should be prohibited as it can evoke interference in the intraneural blood flow [12, 19]. Therefore, setting $\Delta P = 20$ mmHg in Eq. (4.1) derives a critical Δr at which the cuff electrode can be installed without inhibiting the blood flow of the nerve. The MENCE has an internal cuff radius of 1.27 mm with a wall thickness of 50 μm, Young's modulus and Poisson's ratio of 2.5 and 0.4, respectively (as the Poisson's ratios of Vecstar CT-Z and CT-F are not provided, that of the Sumikasuper E5000 is used instead). In addition, weight of the MENCE including two magnets should be also taken into consideration. The total weight of the MENCE is 0.873 mN, with 0.275 mN accounted for by the magnets. By taking the weight of the MENCE into consideration, 0.873 mN of force is placed on the 28.26 mm^2 area of a half cylinder. Then ΔP value, the pressure exerted on the nerve, becomes 19.77 mmHg to calculate the critical CNR. With the given values, Δr becomes 0.351 mm, establishing a critical CNR of 0.783. This means it is safe to design the cuff radius 21.7% smaller than that of the nerve.

4.1.2 Simulation Parameters

The current amplitude of 0.106 mA was used in the in vivo assessment. The current stimuli were strong enough to evoke the response, therefore we also attempted to find a threshold level. The stimulation level was swept while observing the extent of tongue extrusion of the animal. As a result, the threshold level was found to be 41.0 μA at a rate of 0.635 kHz with 0.375 ms pulse duration.

As the stimulation level increased, the tongue response became too strong in that the stimulation is to be performed during sleep. Figure 4.1 shows the EMG signals on the chin measured using Embletta × 100 (Natus Neurology Inc., USA) at the stimulation level of 41.0 μA (top) and 53.2 μA (bottom). The EMG signal with the lower stimulation level showed a single peak in one period while double peaks were observed at the higher stimulation level. The first common peak represents the contraction while the second peak represents relaxation of the muscle. This second peak can be used as a indicator whether the stimulation level is excessive.

4.1.3 Extension to Multi-channels

Although only two electrodes are used for bipolar stimulation here, our proposed closing method can be also extended to multi-electrode cuff, such as the one developed by Schuettler et al. [2]. We may simply use a long magnet in the cuff or we may construct the cuff with twelve single-channel cuffs each with a short magnet. In the latter construct, individual electrode of the cuff can be easily installed or removed by taking advantage of our proposed closing method.

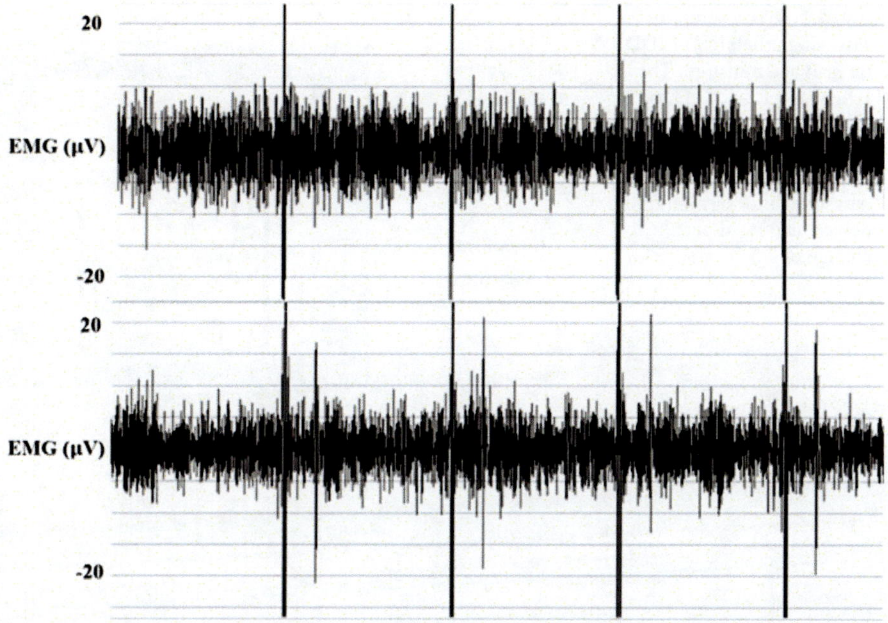

Fig. 4.1 The measured EMG data at current level of 41.0 μA (top), and 53.2 μA (bottom)

4.2 Palatal Implant System for SPS

4.2.1 Reliability of Intra-Oral Device

The intra-oral device (IOD) is a device which is to be placed in the oral cavity. Though the device is encapsulated by silicone elastomer, there is a possibility for the device to fail due to penetration of saliva. Furthermore, heat generated from the power amplifier in the device may result in burn of oral mucosa. Therefore, an assessment was conducted to evaluate thermal and hygric reliability of the device and the setup of the assessment is shown in Fig. 4.2a. The IOD was left in a convention oven for enough time until the temperature of the solution reaches 38 °C. Temperature of the solution near the device was measured at every 30 min, checking whether functions of the device were working well. Figure 4.2b shows the result. The IOD operated for 6 h and the maximum increasement of temperature was 2.5 °C at the time of five hours from the start. After the five-hour, the temperature started to decrease and this seems to be due to lowered power supply from the battery. From the result, the reliability of IOD on the head and moisture, was demonstrated.

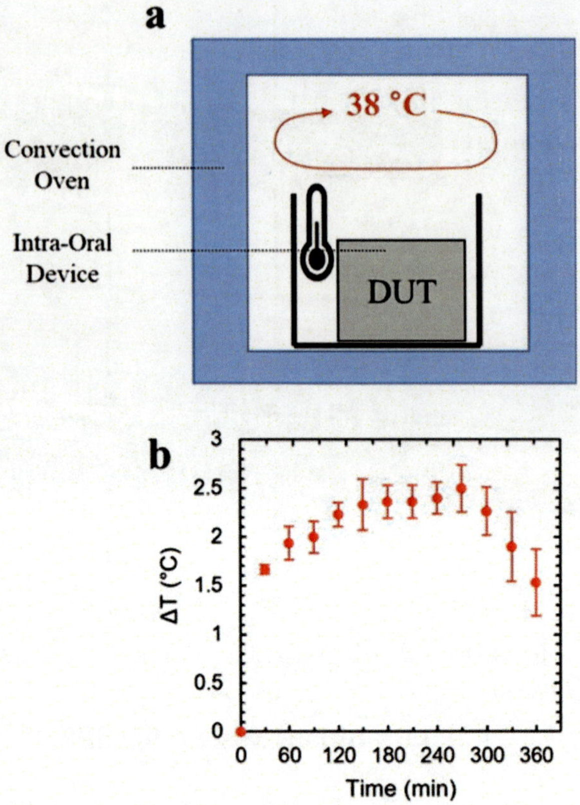

Fig. 4.2 **a** Test setup for assessing reliability of IOD on heat and moisture. The IOD was left in 38 °C for enough time. **b** Temperature variance according to time. Temperature rise was measured after turn-on of the device, and rated peak change of 2.5 °C after 5 h

4.2.2 Palatoglossus Coupling

From the accumulated clinical experiments of the hypoglossal nerve stimulation, retropalatal opening of OSA patients has been detected by researchers, although hypoglossal nerve stimulation targets retrolingual opening accomplished by contraction of the genioglossus [3, 4]. The unique phenomenon of palatoglossus coupling can be explained by presence of the palatoglossus, which is the muscle that arises from the soft palate and extend to the sides of the tongue.

The couplings were observed through experiments conducted in this work not only in the hypoglossal nerve stimulation but also, interestingly, in the soft palate stimulation. Figure 4.3a shows the effect of HNS on the soft palate. The clear contraction of the soft palate was observed with contraction of the genioglossus, simultaneously. On the other hand, contraction of the genioglossus was shown during the soft palate stimulation, as shown in Fig. 4.3b. This indicates that the soft palate and the genioglossus are cross-linked and shows the possibility of multi-region opening through upper-airway stimulation.

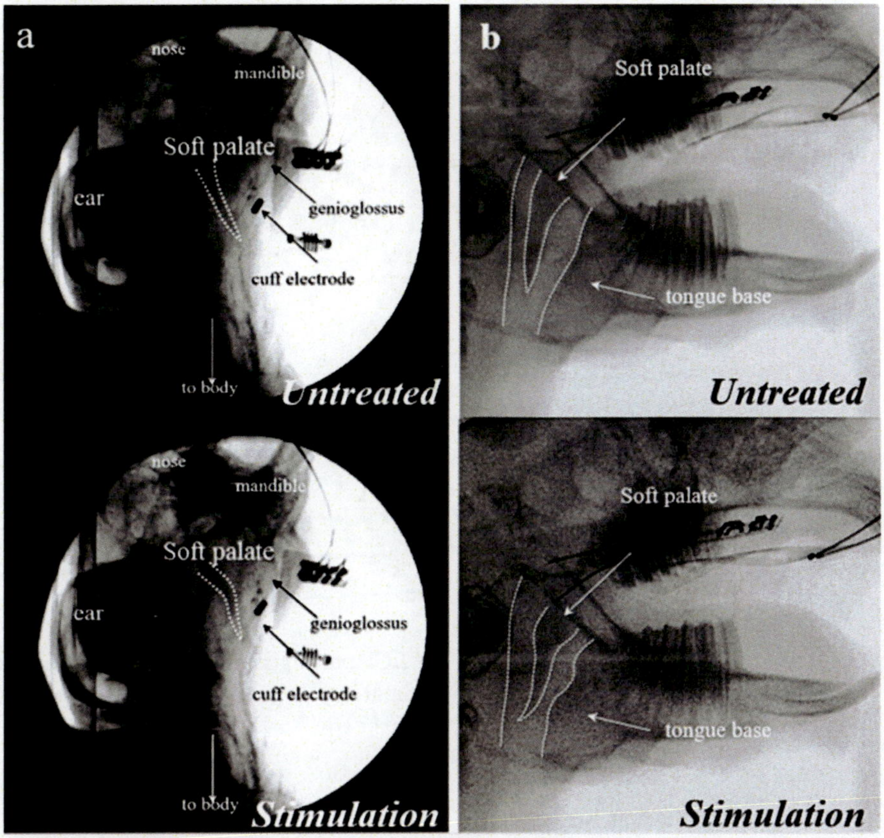

Fig. 4.3 Palatoglossus coupling shown in in vivo assessments **a** contraction of soft palate during HNS (top: untreated, bottom: stimulated) **b** contraction of genioglossus during SPS

4.2.3 Potential Developments

4.2.3.1 Monolithic Packaging of Implant

The implant was fabricated on LCP substrates of which the biocompatibility and long-term reliability have been repeatedly demonstrated in previous studies [5–10]. However, the implant was packaged by a silicone elastomer, which is not suitable for long-term implantation due to its relatively high water absorption rate. Therefore, it is desirable to package the implant with a material with a low water absorption rate, and LCP can play such a role due to its chemical inertness and extreme water absorption rate [10, 11]. Furthermore, an additional benefit of the LCP packaging comes from a monolithic package, which can prevent problems that may otherwise

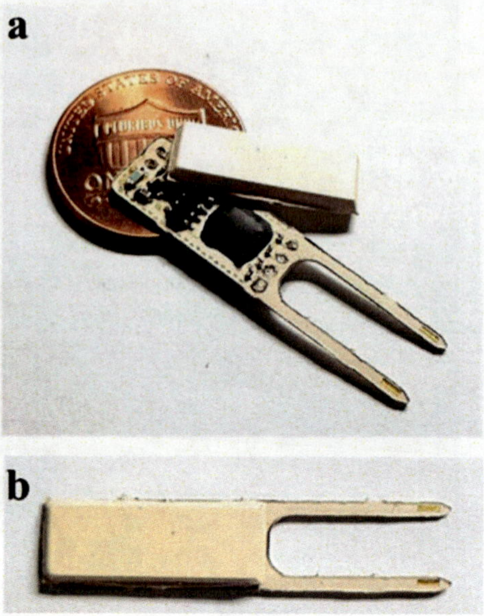

Fig. 4.4 Suggested monolithic device packaged by LCP. **a** Separated parts of the lid and the body, **b** Mock-up of the integrated device

occur due to seams at hetero-junctions. The LCP packaging method is well described in the reference and can be applied to the palatal implant, as shown in Fig. 4.4, which presents a mock-up of the suggested package [12].

4.2.3.2 Integration of Sensors

The intra-oral device, shaped in a mouth-piece, can be synergetic if various sensors of temperature, pressure, and humidity are embedded for breath detection. With the acquired data, the sleep patterns of OSA patients can be tracked and the data can be utilized for the quantitative evaluation of sleep quality using the apnea-hypopnea (AHI) index. Furthermore, the signal synchronized to the breath pattern can contribute to constructing a closed-loop system of the palatal implant (Fig. 4.5). By selectively stimulating the soft palate when apnea is detected, fatigue of the muscle as well as the power consumption of the device can be decreased. A polymer pressure sensor made of PDMS and carbon nanotubes was fabricated to detect oral pressure, and a test result of breathing detection was presented in terms of changes in its diaphragm resistance [13]. The resistance change in the diaphragm was sensitive to small perturbations in air pressure, and real-time detection was possible. With the accurate and sensitive breathing sensor, the palatal implant system can be a more efficient and helpful device.

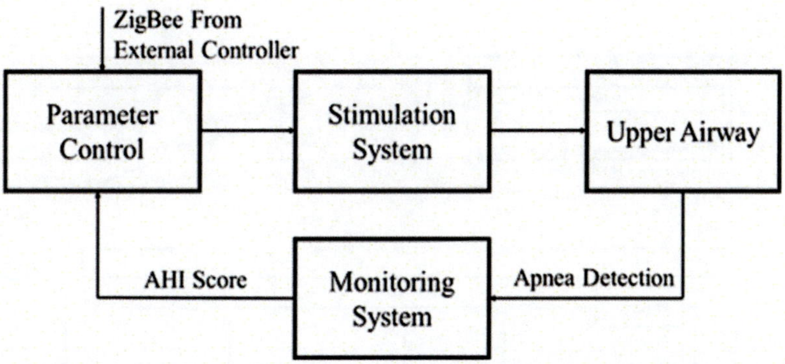

Fig. 4.5 Suggested diagram of the closed-loop palatal implant system

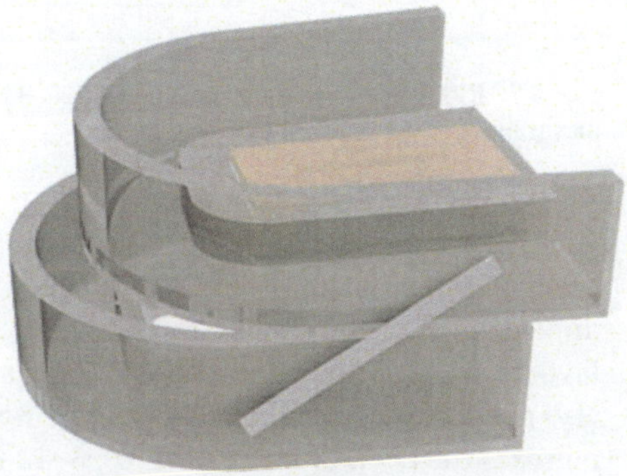

Fig. 4.6 Suggested intra-oral device with mandibular advancement

4.2.3.3 Application of Mandibular Advancement

A mandibular advancement device (MAD) is a treatment apparatus devised for retrolingual opening by advancing the mandible. Its efficacy has been reported to be under 50% in a percentage reduction in AHI with high adherence; therefore, it has been suggested as an alternative to CPAP treatment [14–16]. Having the shape of an oral appliance, an MAD can be applied to the intra-oral device of the palatal implant system, as illustrated in Fig. 4.6. If the two devices are merged, an improved therapeutic effect can be expected by opening both retropalatal and retroligual regions.

4.2.3.4 Improved ASIC Design for Palatal Implant

An application specific integrated circuit (ASIC) is essential to implantable devices. The ASIC used in the experiment successfully serve as a biphasic current generator in the palatal implant. However, some points of the ASIC needed to be modified for an improved palatal implant.

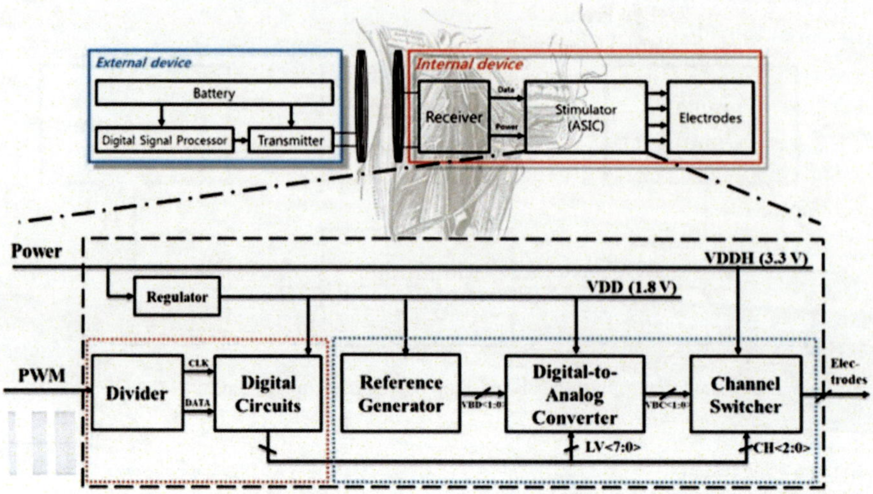

Fig. 4.7 Block diagram of the designed ASIC. The demodulation part (red) and the pulse generator part (blue) are two major blocks of the chip

First, the die size of the ASIC is 3.3 mm × 3.3 mm, which is quite large considering their land patterns and molding epoxies in a wire-bonding process. The large dimension is from unnecessarily many channels for this application and manual layouts of the digital circuit design. A small error margin of the PWM demodulation also needed to be modified. Current demodulation depends on the number of rising pulses at the DATA pin; 3–7 for '0', 8–12 for 'F', and 13–17 for '1', while the others are considered as an error. However, the existence of 'F' takes up unnecessary error margins, which may lead to the possibility of an error in demoulation.

A new ASIC design is proposed to overcome the drawbacks of the existing ASIC in this section. The newly designed ASIC has square dimension of 903.6 μm × 903.6 μm, and it supports eight output channels, including a reference. The dimension could be minimized by reducing the number of channels and using a standard cell-based digital circuit design. Furthermore, a new protocol of the ASIC was designed without 'F' to increase error margins of '1' and '0'.

Circuit Design

The ASIC requires both power and data to generate a biphasic current pulse. The ASIC requires two power sources, one is VDDH (3.3 V) of a higher voltage level, and the other is VDD (1.8 V) of a lower voltage level. VDDH serves as a compliance voltage of the stimulation output, while VDD is the supply for all other power-consuming components.

First, in the demodulation part (red dotted region in Fig. 4.7), the ASIC receives a PWM signal and regenerates a clock (CLK) and a train of data pulses (DATA) from the signal. Then the digital circuit demodulates the data and generates control signals that include information regarding level, frequency, duration, and channel selection.

In the pulse generator part (blue dotted region in Fig. 4.7), there is a temperature-compensated reference current generator, which is less sensitive to temperature variation, referencing a previous study [17]. The reference current is delivered to the DAC through a current mirror. The DAC multiplies the reference current based on the level control signal from the digital circuit. The determined current level is further delivered to a channel switcher consisting of H-bridges of which two channels are selected for a cathode and an anode, respectively.

Digital Protocol

Each bit comprising a word is determined by the number of received pulses in a PWM signal. Unlike the existing ASIC, a bit can be either '1' or '0', not F, which is the bit indicating the end of the word. Rather, in this protocol, the end of each word is identified by counting the number of received pulses. By removing F, counting margins of the bit decision can be more flexibly determined. The number of pulses ranging from three to nine corresponds to a bit '1', twelve to eighteen to '0', and the others to an error. That is, duty cycles of PWM ranging from 15 to 45% correspond to '1', and those ranging from 60 to 90% correspond to '0'.

Eleven bits constitute a word and the word has for modes, namely, (AM_SET_STIM), reset (RST), duration setting (DUR_SET), channel setting (CH_SET), and amplitude setting and stimulation (AM_SET_STIM). To initialize the circuit, enough '0's over twelve should be provided to make sure that the circuit is under a reset state. When the first '1' is detected after the '0's, the circuit initializes a counter that counts the number of input bits to define the end of the word. The second left-most bit determines the state to either setting modes (DUR_SET, CH_SET) or a stimulation mode. The duration and active channels of the stimulation output can be set in the setting modes. In the stimulation mode (AMP_SET_STIM), control signals switching transistors of the H-bridge are printed out according to the duration, active channel, and the amplitude information, which is determined in the stimulation mode. Every word except RST has a parity bit at the right-most to check whether the word is in even-parity. By performing a parity check, the rate of erroneous demodulation can be reduced.

Circuit Implementation

A simple illustration of the pulse-generating circuit is depicted in Fig. 4.8. First, a reference generator creates a constant current level of 10 μA at 27 °C. The reference current is copied to the DAC in terms of the bias voltages of the current mirror. The current is multiplied to eight branches, each of which is controlled by digital signals. The ASIC is capable of generating a current amplitude from 10 μA to 2550 μA with 256 steps. The duration of the current is also controlled by the digital signals.

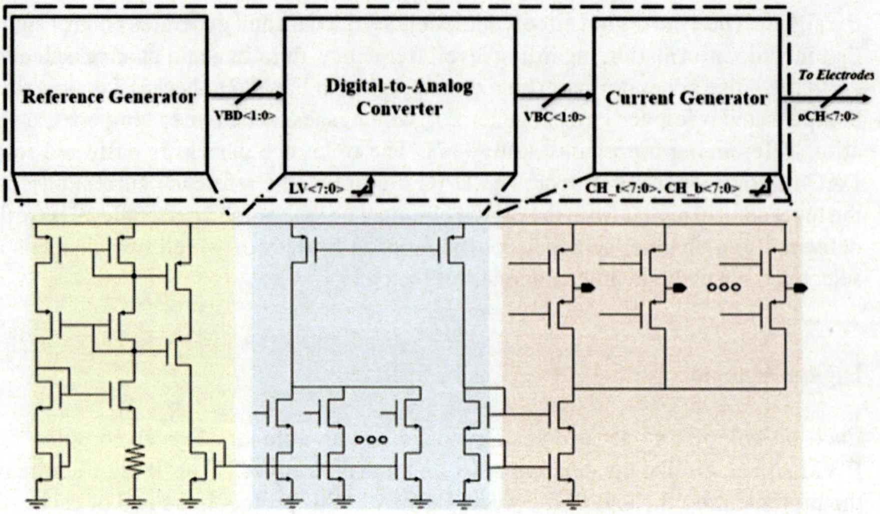

Fig. 4.8 Pulse generating circuit of the designed ASIC. Temperature-compensated current generator is followed by DAC, and the current generator comprised of a series of H-bridges

The duration of the current pulse ranges from 24 to 2028 μs with 128 linear steps of 16 μs, including a resting time of 8 μs. The pulse rate is determined by the rate of the rising edge of the message signal, which is possible up to 125 kHz with a 2.5-MHz carrier frequency. The current amplitude is further mirrored to the current generator in terms of bias voltages. The current generator is supplied by VDDH, the high-voltage supply which serves as a compliance voltage of the current pulse. A series of H-bridges are tied up to the current sink, which is mirrored from the DAC. Therefore, the current output having parameters of the desired amplitude, duration, and frequency is generated through the desired channels. The integrated analog and digital circuit layout is shown in Fig. 4.8. The die size of the chip was 0.90 mm × 0.90 mm, with a 0.30 mm × 0.28 mm digital core. The designed size is significantly smaller than the previous ASIC chip having dimensions of 3.3 mm × 3.3 mm, which constitutes a 53.3% reduction in the die size.

References

1. Naples GG, Mortimer JT, Scheiner A, Sweeney JD (1988) A spiral nerve cuff electrode for peripheral nerve stimulation. IEEE Trans Biomed Eng 35(11):905–916. https://doi.org/10.1109/10.8670
2. Schuettler M, Donaldson N, Seetohul V, Taylor J (2013) Fibre-selective recording from the peripheral nerves of frogs using a multi-electrode cuff. J Neural Eng 10(3):036016
3. Safiruddin F, Vanderveken OM, de Vries N, Maurer JT, Lee K, Ni Q, Strohl KP (2015) Effect of upper-airway stimulation for obstructive sleep apnoea on airway dimensions. Eur Respir J 45(1):129–138. https://doi.org/10.1183/09031936.00059414

4. Heiser C, Edenharter G, Bas M, Wirth M, Hofauer B (2017) Palatoglossus coupling in selective upper airway stimulation. Laryngoscope 127(10):E378–E383. https://doi.org/10.1002/lary.26487
5. Aihara K, Chen MJ, Chen C, Pham A-VH (2012) Reliability of liquid crystal polymer air cavity packaging. IEEE Trans Compon Packag Manuf Technol 2(2):224–230
6. Sundaram V, Sukumaran V, Cato ME, Liu F, Tummala R, Weiland JD, Nasiatka PJ, Tanguay AR (2011) High density electrical interconnections in liquid crystal polymer (LCP) substrates for retinal and neural prosthesis applications. In: 2011 IEEE 61st electronic components and technology conference (ECTC). IEEE, pp 1308–1313
7. Hwang GT, Im D, Lee SE, Lee J, Koo M, Park SY, Kim S, Yang K, Kim SJ, Lee K, Lee KJ (2013) In vivo silicon-based flexible radio frequency integrated circuits monolithically encapsulated with biocompatible liquid crystal polymers. ACS Nano 7(5):4545–4553. https://doi.org/10.1021/nn401246y
8. Gwon TM, Kim JH, Choi GJ, Kim SJ (2016) Mechanical interlocking to improve metal–polymer adhesion in polymer-based neural electrodes and its impact on device reliability. J Mater Sci 51(14):6897–6912
9. Lee SW, Min KS, Jeong J, Kim J, Kim SJ (2011) Monolithic encapsulation of implantable neuroprosthetic devices using liquid crystal polymers. IEEE Trans Biomed Eng 58(8):2255–2263
10. Jeong J, Hyun Bae S, Seo JM, Chung H, June Kim S (2016) Long-term evaluation of a liquid crystal polymer (LCP)-based retinal prosthesis. J Neural Eng 13(2):025004. https://doi.org/10.1088/1741-2560/13/2/025004
11. Gwon TM, Kim C, Shin S, Park JH, Kim JH, Kim SJ (2016) Liquid crystal polymer (LCP)-based neural prosthetic devices. Biomed Eng Lett 6(3):148–163
12. Jeong J, Bae SH, Min KS, Seo J-M, Chung H, Kim SJ (2015) A miniaturized, eye-conformable, and long-term reliable retinal prosthesis using monolithic fabrication of liquid crystal polymer (LCP). IEEE Trans Biomed Eng 62(3):982–989
13. Tallis H, da Costa SJK, Jeong-Whun K, Jungmin S, Jin-Woo C (2018) Breathing detection using a polymer pressure sensor. paper presented at the SMIT-IBEC, Seoul, Korea, Nov 2018
14. Brown EC, Cheng S, McKenzie DK, Butler JE, Gandevia SC, Bilston LE (2013) Respiratory movement of upper airway tissue in obstructive sleep apnea. Sleep 36(7):1069–1076. https://doi.org/10.5665/sleep.2812
15. Chan AS, Sutherland K, Schwab RJ, Zeng B, Petocz P, Lee RW, Darendeliler MA, Cistulli PA (2010) The effect of mandibular advancement on upper airway structure in obstructive sleep apnoea. Thorax 65(8):726–732. https://doi.org/10.1136/thx.2009.131094
16. Cobo J, de Carlos F, Suarez A (2009) Mandibular advancement device (MAD®) to treat sleet apnoea-hypopnoea syndrome and chronic snoring. In: 25th southern biomedical engineering conference 2009, 15–17 May 2009, Miami, Florida, USA. Springer, pp 349–354
17. Fiori F, Crovetti PS (2005) A new compact temperature-compensated CMOS current reference. IEEE Trans Circuits Syst II: Express Briefs 52(11):724–728

Chapter 5
Conclusion

The significance as well as a summary of the thesis are provided in Conclusion.

In this thesis, neural prostheses were suggested as novel approaches to treat obstructive sleep apnea. Two prosthetic devices were designed, developed, and demonstrated for tongue-related and palate-related OSA patients by opening the retrolingual and the retropalatal regions, respectively.

First, as a method for retrolingual opening, a magnetic cuff system for hypoglossal nerve stimulation was designed and fabricated. The cuff electrode featuring magnetic closure allows repeated and reliable reinstallation, which is a very helpful feature to identify the correct nerve. A pair of rare-earth magnets were embedded in a liquid crystal polymer (LCP) substrate of which the water absorption rate is very low compared to other implantable polymers, such as polyimide, parylene-C, and PDMS. A thin gold layer (10 μm) was patterned on the LCP substrate using a series of semiconductor processes. Seed layers of Ti and Au were deposited on the LCP using an E-gun evaporator and then electroplated and etched for patterning the metal layer. Further thermal lamination and laser machining processes were performed to embed the magnets and complete the fabrication. The fabricated electrode had an electrode site area of 3.5 mm × 1.6 mm with a substrate thickness of 50 μm.

The fabricated cuff electrodes were evaluated both in vivo and in vitro to demonstrate its feasibility for hypoglossal nerve stimulation. As in vitro evaluations, electrochemical impedance spectroscopy (EIS) and cyclic voltammetry (CV) were conducted using a three-electrode system in a phosphate buffer solution (PBS). From the EIS result, the magnitude of the impedance of the electrode at 1 kHz was 638 Ω with a phase angle of −67.8°, which is many times lower than that of conventional microelectrodes due to its large surface area. The cathodic charge storage capacity (CSC_c) was calculated from the CV plot to be 1.69 mC/cm^2, which is large enough compared to the required amount of charge for our stimulation, 7.68 nC. A seven-month-old New Zealand white rabbit was used as a subject of the in vivo evaluation. The cuff electrode was installed on the hypoglossal nerve of the rabbit and percutaneously connected to an external pulse generator. The stimulation current pulse had

J. Seo, *Development of Implantable Electronics as Novel Approaches to Obstructive Sleep Apnea*, Springer Theses, https://doi.org/10.1007/978-981-15-8327-8_5

parameters of 41.0 μA amplitude, 635 Hz pulse rate, and 375 μs duration. Snapshots of the upper airway of the subject were obtained using C-arm fluoroscopy in both the presence and absence of stimulation. Protrusion of the tongue could be visually observed during the stimulation, synchronized to the applied pulse. Contraction of the genioglossus was also observed via C-ram fluoroscopy snapshots.

A novel palatal implant system was proposed for retropalatal opening. The system consists of an implant that delivers current pulse to the soft palate and an intra-oral device (IOD) that transmits power and data to the implant via an inductive link. In the implant, a multi-layered LCP laminate, encapsulated by PDMS, includes coil, electrode, and circuit layers, generating and delivering a biphasic current pulse for soft palate stimulation. The circuit regenerates power and data for the pulse generator ASIC by rectifying and demodulating received RF signal, which is transmitted from the IOD. The RF signal is generated by a class-E amplifier in the IOD with an attached ASK modulator, which inputs the data signal received via a ZigBee receiver. All the components of the IOD including a battery, are stacked in a 30 mm × 15 mm × 10 mm cuboid, which is placed in the cavity of an oral appliance.

Evaluations on the inductive link were performed to measure the power transmission and the reliability of the data telemetry. The received peak-to-peak voltage was 10 V, and the total power efficiency was 12.4% at the telemetry distance of 5 mm. The duty-cycle errors of the demodulated PWM were measured at distances ranging from 2 to 8 mm, and the errors were all allowable within the error margin of 10%.

In vitro evaluations were conducted to measure and compare the electrochemical impedance spectrum and cathodic charge storage capacity of Au and EIROF (electrodeposited iridium oxide film) electrodes. The magnitudes of the impedance and CSC_c of EIROF were hundred of times lower than those of Au. The voltage transient was also measured to ensure safe stimulation, and the charge injection capacity of EIROF did not exceed 0.16 mC/cm^2, which is a far smaller amount of charge than our stimulation parameter.

To demonstrate the feasibility of the system for soft palate stimulation, an in vivo test was conducted using a rabbit. The implant was implanted into the palate of the rabbit, and power and data were received from a transmitter coil through the palatal skin. A biphasic current pulse, having parameters of 1.00 mA amplitude, 100 Hz pulse rate, and 540 μs duration, was applied to the soft palate. As a result of the stimulation, contractions of the palatoglossus muscle as well as the soft palate could be observed via C-arm fluoroscopy.

In the discussion, possible developments of the palatal implant system were proposed, such as sensor integration and an application of a mandibular advancement device. Finally, the design of a mixed-signal ASIC chip for minimized die-size and improved communication protocol was suggested.

Sleep apnea is becoming a prevalent disorder in modern society. Still, there are patients who do not benefit from the existing therapies. Upper airway stimulation can play an important role in treating OSA considering that airway patency is closely related to activation of the pharyngeal muscles. We hope that the research reported in this thesis will become a stepping stone for the development of a breakthrough therapy that will help patients who suffer from OSA.

Author Biography

Dr. Jungmin Seo graduated Seoul National University, Korea and received his Ph.D. degree in the same university. He researched neural/neuro-musclar stimulation in the Nano-Bioelectronics and Systems Laboratory. He currently works for Samsung Electronics as a staff engineer.

J. Seo, *Development of Implantable Electronics as Novel Approaches to Obstructive Sleep Apnea*, Springer Theses, https://doi.org/10.1007/978-981-15-8327-8